A Farmer's Almanac

Stories about Land, Food, and Life
—*Fascination of Discovery*—

A Farmer's Almanac

Stories about Land, Food, and Life

—*Fascination of Discovery*—

Volume I
2014–2018
Part 1

by Drausin Wulsin

Copyright © Drausin Wulsin, 2019
Second Printing Copyright © 2023
ISBN: 978-0-578-54760-2
Red Stone Farm, LLC
Hillsboro, Ohio
www.redstonefarm.org
All Rights Reserved

Design & Production by Jennie Hefren
jshdesign98@gmail.com
Editorial Direction and Project
Management by Howard Wells
hiwsandwell@gmail.com

Printed in the United States of America

*To Julia Child and Allan Savory,
who taught one of us how to cook
and the other of us how to think.*

Preface

This is the second edition of Volume I. When first composed, I didn't know there would be three more volumes to come. But time did its deed, so now we have a series of stories to convey covering the course of our compelling journey producing nutrient-dense food in a sustainable fashion. These stories are organized as follows:

Volume I, Part 1
2014 – 2018
Fascination of Discovery
Color of Brown

Volume II
2019 – 2020
Wonder of Actualization
Color of Green

Volume III
2021 – 2023
Virtue of Adaptation
Color of Blue

We wanted to keep the chronology of events moving for the reader, and thus divided the early years in half. If you are still with us after the first three volumes, we then return you to our earlier years with more rich, timeless, untold stories from 2014 – 2018.

Volume IV – Volume I, Part 2
2014 – 2018
Fascination of Discovery
Color of Brown

DFW

Contents

Introduction 10
1. A Dying Cow, Two Noble Women, & a Finite Farmer 13
2. Heritage Table 15
3. Grazing Tall 17
4. Ribeyes & Lovemaking 20
5. Rest, Death & the Tree of Life 22
6. Field of Dreams 24
7. A Mother's Love 26
8. Cattle Auction 29
9. Shimmering Spires 33
10. Digging Deep 36
11. Artisanal Logging 39
12. Little House on the Diary 42
13. Beauty of Fat 44
14. Pulling Posts 47
15. Land Stewardship 50
16. Liberty Bell 52
17. Magical Discourse 54
18. Dancing Shadows 59
19. Country Music 61
20. The Tribe 64
21. Strands of Life 66
22. Animal Impact 68

23. Colors of Redemption 70
24. Ancient Roots 73
25. Dignity of Labor 76
26. Savory in Georgia 79
27. Depth of Heart 82
28. Strength of Mind 85
29. Power of Soul 88
30. Beauty of Spirit 91
31. Table Travel 94
32. A Day's Movement 96
33. April Action 98
34. Tragedy & Hope in Pike County 101
35. Power of Water 105
36. Ancient Wisdom 110
37. Power of Carbon 114
38. Goddess Energy 118
39. Wedding Altar 122
40. Biodiversity & Humility 124
41. Wedding Celebration 128
42. Collecting Seeds 132
43. Wildfire 135
44. Buildings 137
45. Slow Fat 140
46. Man's Best Friend 142
47. Flood Waters 146
48. What's in a Name? 149
49. Walk in the Woods 152
50. Marketplace 155
51. Turkey Tail 159
52. Summer Solstice 162
53. Tall Grass 165

54. The Long Road	168
55. Soles of Our Feet	171
56. Lyrical Lyra	175
57. Food & Politics	178
58. Pursuing Excellence	181
59. Virtue of Travel	184
60. Language & Grassfed Meats	187
61. Breeding Season	190
62. Water Worries	192
63. Fearless Cook	198
64. Last Snow	201
65. Rain Blast	203
66. Side by Side	205
67. Spirit Quest	208
68. House of Love	211
69. Moving Dirt	215
70. Rainbow of Life	219
71. Coyote Blossom	222
Afterword	225
Photo Credits	227
Index	228

Introduction

The original intent of writing these newsletters was to improve on-line orders and sales. Children and nephews had advised we had better employ social media to tell our story or forever be lost to history. So, heeding their precocious but sound advice, I have written stories to share news of this farm nearly every week for the past five years. Initially, it was daunting to find a story worth telling on a regular basis. After a while, however, I realized the story did not need finding, because it sat before my eyes, if they were open. We have been on such a rapid learning curve the past five years, that there has been much to illuminate.

The challenge upon which Susan, my partner and wife, and I are embarked, direct marketing of grassfed foods, is immense. We are unintentionally opposing the industrial food system, which has evolved over 100 years to a point that it now holds an ironclad grip on our culture, through low pricing and high convenience. It was fortuitous we didn't know at the outset exactly how large the challenge would be to face this Goliath or we might not have undertaken it. But in so doing, we have stretched ourselves to limits previously unreached, discovering new inner strength and a few cracks in the armor of the giant.

When heading into the unknown, as we have, one is essentially calling for spiritual guidance. We did not anticipate this. The depth of the experience that has unfolded, with many striking failures followed by slowly growing successes, has been beyond our ability to imagine. For instance, in our first year, we came upon a pasture of 40 dead lambs, due to parasites, which reflected poor management on my part. That was a clear low point. Four years later, we sold 100 grilled and frozen sliders per hour for four hours, at the farmers market, due to Susan's expertise, which was a high point. Throughout the oscillations, we have held basic confidence in the promise of the mission and in our combined talents.

It began to dawn on us we were on a quest for spirituality in combination with a quest for a sustainable business. This experience has generated uncommon strength to persist and has made us feel ever more connected to the humanity of our customers, for each one of us is on a spirit quest, no matter where we live or what we do.

Writing these stories has helped us explore and refine our philosophy of life and business, at the center of which stand three values: health, taste, and connection. Health arises from the land, taste from the food grown on it and prepared in the kitchen, and connection from the financial, emotional, and philosophical bond with customers and partners.

This journey started thirty years ago, when we launched one of the first grass-based organic dairies in Ohio, which took twenty years to become profitable. Twenty years ago, we began the long process of developing a wetlands mitigation bank, which took fifteen years to become profitable. Ten years ago, Susan and I birthed Grassroots Graziers, aka Grassroots Farm & Foods, to produce and market grassfed foods. In the past six years, we began direct-marketing and five years ago I began writing stories about the journey. I am not sure the stories have contributed to sales in the short run, so a while ago I stopped worrying about that, and just concentrated on the story. We know the stories have contributed to connection, because of many comments received. With sales currently rising, however, we ponder the correlation. In the meantime, enhanced connection is complete nourishment of its own.

One is never alone on long journeys. We have been primarily steered and comforted by our two mentors—Julia Child and Allan Savory. Susan began teaching herself to cook as a hungry ten-year-old. Julia Child became her guiding force, to whom she has largely deferred throughout a career of culinary scholarship. As you will see in the pages ahead, the results of this connection are extraordinary. Susan pays significant homage to Julia almost seven nights a week.

Allan Savory uniquely articulates how to make sustainable decisions within the great chaos circulating at the intersection of land, money, and people. Without his profound insights and guidance, we could not have wrought the productive changes to this farm we have over the past thirty years. His presentation of Holistic Management is winning, as it suits all levels of engagement, from merely intuitive consideration to deep analysis. I constantly hear him in my head, as I drive around our farm trying to make sense of the movement. He has brought us a measure of abundance and a sense of promise. We are truly indebted to him. One reason to publish these stories now is so he and Jody, his partner and wife, may receive this tribute, while still in their golden years.

Other partners have been dedicated employees, who have kept us from submerging along the way: Bob, Whitney, Brendan, Beth, Cole,

Kathy, and now, Clark, Mike, and Stephani. Sebastien, Emma, and Alex have contributed at the marketplace and in the kitchen. Without the considerable efforts of these fine people, we would not be here to tell the tale, and we extend special gratitude to them.

The stories that follow are a sampling of many that were written. May they bring you a taste of our spirit quest, on this beautiful land, while engaged with so many supportive customers and partners.

A Dying Cow, Two Noble Women, and a Finite Farmer

February 19, 2014

    The intention was to leave our house in Batavia this morning to drop meat off at Helene's and visit my father, but my cellphone apprised that one of our cows was "down" at the farm. An hour later I found Red Devon W92 on her side in labored breathing. Cole and Whitney were on the morning shift, and had done their best to right her, but she resisted. I then struggled for three more hours with tractor, ropes, and bales of straw to bring her to an upright position, so she could breath, which met with mixed success. The vet showed up in the fourth hour, and together we righted her another time. He applied an IV of magnesium and calcium but did not leave feeling optimistic. By the time his taillights disappeared, she had gone prone again. Into the fifth hour of hugging, pushing, and pulling, with her manure and saliva all over my front, I tried again to bring her upright and buttress her with large round bales on each side. But other cows were crowding around, pushing the bales askew, and licking her... Something was clearly just not right.

So, I gradually began to submit. Both she and I were exhausted by this time. Since her will to live seemed to have dissipated, I then thought the kindest thing might be to put her down, so she wouldn't suffer through the night. A former employee had stolen my gun, so I went down the road to Landis' to borrow his. He had no slugs for his gun but returned with me to help load the noble cow onto the front-end loader so we could move her to the shelter of the barn. I called another neighbor for a rifle, but she wasn't home. That plan wasn't working either.

After a day of struggle and best intentions, all I was able to accomplish was moving her into the barn. Her kind and gentle eyes kept surveying and boring into my growing ineptitude. After a while, I began to feel it wasn't me she was looking at, but something through and beyond my poor purpose. I finally realized I made no difference at all in this situation. So, I put the tractor and bales of straw away, laid my hand on her great shoulder, and in heart-heavy deference, began my retreat to Batavia to have dinner with Susan. During the drive, I tried to make sense of the intimate smells of her body upon mine, the wisdom in her eyes, and her increasingly labored breathing.

Upon entering the house, Emmylou Harris was intoning Appalachian minor keys from one wall to another and the aroma of lamb stew emanated from the kitchen. A tall goddess embraced me, with empathy flowing. By the time I was out of the shower and on the couch with a beer, Pavarotti was ushering his resonate tenor-storm from bowels of the earth to heights of mountains. So, in compliance, I let myself cry—over the tenor storm, the noble cow, the noble wife, and how finite this farmer is...

When we sat down for dinner, YoYo Ma was transporting, and the richness of the day only grew. Lamb stew arrived with fresh carrots, apricots, almonds, and parsley—another touch of exultation.

How does one make sense of all of this? I don't really know.

But tomorrow I will take that dear cow to the woods and try to listen one more time to what she has to say. She and you and we are in this great circle together, and how rich life is as a result.

Heritage Table

April 26, 2014

This wedding gift to my son, made over the course of the winter, and delivered over Easter weekend to his abode in Manhattan, has led me to ponder the meaning of a table.
What happens around and upon a table?

- Nutrition is administered,
- Newspapers are read,
- Homework is conducted,
- Groceries are stacked,
- Artful food is presented,
- Delicious meals are savored,
- Stories are recounted,
- Community is gathered,

- Values are taught,
- Emotions are shared,
- Concepts are pondered,
- Rest is taken,
- Love is offered.

All of this renders the eating table a potent place.

This presumes one, in fact, sits at a table. Many cultures employ a different surface for these functions, which is the great earth itself. That indeed is the ultimate table, and would be the ultimate wedding gift, but I couldn't pull that off.

The wood came from 100-year-old poplar milled for a granary on our farm. We took down the sagging granary twenty years ago, but saved the boards from the walls, which are 18 inches wide—of greater width than can be found in trees today.

Converting the worn, dusty boards into surfaces fit for worship was a journey in itself—much sanding, endless scrubbing with steel wool, and many coats of varnish, until the grain of the wood races forward and sings. The final product is replete with: nail holes, mouse gnawings, cracks, and knots—a completely imperfect version of a finished surface and so perfect.

Why do many of us engage in much labor and activity surrounding food? At minimum, it has to do with bodily nutrition, but what drives us must be something more, like the list above and the feeding of soul and spirit.

Grazing Tall

June 5, 2014

Like a buffalo herd. Grazing tall: high-density, short-duration impact.

We just moved this herd, and these cows and calves will be on half-an-acre of ground for several hours, mimicking a herd of wild ruminants moving across a landscape. The forage is tall, creating lots of quantity, quality, and diversity of nutrition. At the same time, the cows can't wander around looking for the best, so they eat a cross-section of what

is before them and trample the rest. This benefits both animal and soil. The herd is moved forward every 2–3 hours during daylight.

The pasture pictured on the previous page, is the salad bar which our cows are grazing, where the stand of feed is at least three feet tall. Note the wide variety of species—fescue, orchard grass, switch grass, Indian grass, big bluestem, red clover, white clover, and numerous forbs. Sixty head of livestock will deliver over 100,000 pounds of pressure per acre to the land while grazing, enhancing organic matter in the process.

The next picture, by contrast, is of cows grazing on a different farm. About six cows have been in that field for over six months, and the stand of grass is about 2 inches short. Short plants provide only protein for the grazer and no energy, which becomes a nutritional problem over time, solved by feeding corn. The difference in approach to feeding ruminants is noteworthy.

This is a patch of vetch, in bloom. Vetch is a perennial legume, which sheep and cows love. We will be grazing this field in several weeks.

We enjoyed a fabulous Farm Tour this past Saturday. About 35 people were able to attend. The weather was glorious, and everything fell into place. We looked at handling facilities for beef and sheep, witnessed border collies herding the flock of ewes and lambs, discovered the water system was malfunctioning, boarded tractor-drawn wagons and visited the cow herd, watched them move into shoulder-high grass, went down the laneway to meet Landis Weaver to see his calves and cows, and hear about grass-dairying, and then returned for a lunch of grassfed Moroccan burgers, soulful potato salad, local asparagus, and fresh strawberries & whipped cream over ginger pound cake. It was a fun event with great discussion, great food, and the forming of new connections among all of us.

The most important connection was made between those attending and the land itself. All of you are stewards of this farm, joining Susan and me. When you purchase our meat, you support this land and become one of its caretakers. This is invaluable, as land responds quickly to caretaking. We are a community, employing the scarce currency of affection for land, animals, and food to advance our lot and the whole to which we belong.

This is our stone circle, where Susan and I spent this past Saturday evening with a jug of water and a bottle of wine, absorbing the exhilaration of the day and the wonder behind the demanding and powerful journey we are on.

Ribeyes & Lovemaking

June 26, 2014

Do these pictures make your mouth water? On-farm research suggests grassfed ribeye steaks are good for Love.

Universities have not caught wind of this phenomenon yet, but anecdotal evidence keeps surfacing in the country that grassfed ribeyes enhance the making of love, in the same vein as fish and oysters. So, we no longer need to feel disadvantaged we don't live near oceans, as Midwestern pastures can supply all the amour we need.

This raises the question: What is the making of love?

Susan and I pondered this, as we beheld our **pastoral** view over a glass of wine. The longer we looked, the more we saw, and the more our hearts filled. Isn't that what happens when love is being created? It somehow grows from the detail of life into an expanded and unexpected dimension.

This view, framed by stout walnut trees, became deeper and deeper as we studied it and as our stillness grew. The broccoli-studded hillside became more remarkable. Upon closer inspection, we espied dairy cows on the farthest strip of pasture below the broccoli, and below that a sliver of hay just made by Landis and horses. Then the pastures, mowed at various increments, followed by a corral, in which young dairy calves had recently been moved before their journey onto grass. And to the left

are draft horses, replenishing themselves after a full day of work. Each detail is remarkable, but it is the interaction among them that creates a greater story, which fills the heart.

So, as we connect with each other, by body, soul, spirit, and mind, creating wholes greater than sums of parts, and propagating love for which we hunger, remember that grassfed ribeye steaks will take you wherever you need to go. It is being proven in the countryside.

While tough love is a necessary part of life, tough steaks are not. Grassfed beef steaks and lamb chops should not be cooked beyond medium-rare. If you do, they will be tough, and you and we will be disappointed. If you like steaks cooked medium or more, don't invest in dear grassfed products. Ideally, grassfed chops and steaks are cooked rare, leaving you with tender food to match tender love at the table.

Rest, Death & the Tree of Life
July 29, 2014

Beaches of Georgia – offering beauty and rest.

Rest is integral to the workings of nature. Without it, there would be no seasons, dormancy of plants, or hibernation of animals. Natural systems would become exhausted and collapse. At our farm, even during the growing season, we employ rest of 60–90 days to replenish roots of grazing plants, so they re-establish vigor for lean times, like the next grazing, droughts, or winter.

And so it is with people. We need rest as well, which is difficult to achieve on farms and in households, where demand for stewardship does not cease. But vacations and weekends off can help meet the need. Susan and I took four days on the coast of Georgia this past week to find stillness and the strength it provides.

The ultimate rest, of course, is death, which is not uncommon on farms. But it is always replaced by life, keeping the cycles of abundance and faith moving forward. In the past six weeks, both Susan's mother and my father have passed away, after long, celebrated lives. They departed suddenly, though not unexpectedly, leaving behind unique legacies. It is poignant to lose one's elders, but their contribution to the tree of life provides oaken branches against which to lean long after

their departure. And so we carry on, enriched by death, inspired by life, while needing rest to punctuate the journey.

What does rest have to do with food? Well, for chronically busy people like farmers and mothers, dinnertime can provide an important moment of rest. It is when body and soul can intersect to restore on a daily basis. A lovingly prepared meal presented upon a table asks us to receive it with pause, in respect for the provider, the food itself, and our own selves. This pausing is beneficial and is one reason good food is so central to enduring cultures.

Restfully yours.

Field of Dreams

September 4, 2014

*Field of Dreams and Field of Swamp Marigolds
Six feet deep and as far as the eye can see...*

We spend a lot of time managing fields of dreams and of grass, but this one, in our wetlands, we don't have to manage at all. It takes care of itself, and in so doing, takes care of us, for which phenomenon we are increasingly grateful.

In contrast, this pasture is grazed by steers-being-finished-for-harvest and recently-weaned yearlings. We are not forcing them to trample as much as we do the cows, favoring growth of animal over conditioning of

soil and pasture. This field has not been mowed and thus shows desiccated plants, but much green growth lies beneath which the stock quickly find.

The morning parade! One hundred and fifty newly-arrived baby chicks being reared by our neighbor, Pheryl Zimmerman, to our specifications for you. Note the rounded siding, propane heater, and clean shavings —all important management practices. The chicks will be transferred to grass in about three weeks and within eight weeks will be ready to be harvested for freezers. These steps reflect intensive stewardship to produce this superior grassfed product, which can't be found at supermarkets.

A Mother's Love

September 10, 2014

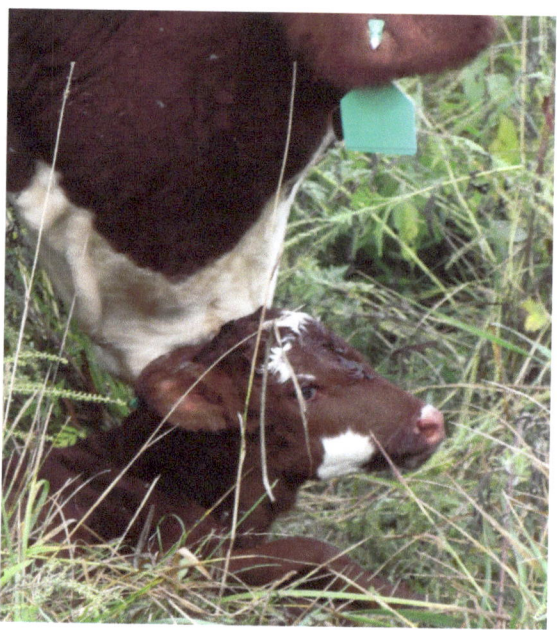

A mother's love is the greatest force in Nature, stronger than hurricanes, earthquakes, or drought.

Calving has begun on the farm, and the magnificent power of motherly love is being demonstrated yet again. The birthing and rearing of young by maternal force has been underway since the advent of time. It symbolizes belief in the future, as does the rearing of our own children and grandchildren. It will not be denied, which is one reason life is good.

To bring you up to date on Frostbite, our aspiring bull-calf, you may remember he was born on the coldest day of the year in late January and is therefore almost nine months old. He is still with his dedicated mother and will be for another month to maximize his intake of maternal

butterfat, to enhance his fertility. The two are in isolation so he doesn't breed others prematurely. It looks like he is on track to becoming a sound breeding bull, but only time will tell.

As the fall equinox approaches, it appears this is a bountiful year for mast and fruit, as demonstrated by heavily laden branches of walnut and pear trees.

This cloud burst of fire greeted me early this morning as I pulled into the barnyard. Its momentary and awesome presence seemed to be reinforcing the passion required to bring food from our land to the marketplace for caring customers like yourselves. When beholding such display, one can't help feeling that God is nearby.

As the yearly ritual of the football season dawns, to touch our primal selves, remember that we offer Moroccan Sliders and Susan's Soulful Chili as perfect accents to your gatherings.

In honor of Motherly Love that brings us forth and the fire that keeps us all advancing, we look forward to connecting soon.

Cattle Auction

October 1, 2014

Red Devons at Brookview Farm, Winchester, Kentucky

This past weekend, Susan and I attended the annual meeting of the Red Devon Association, held in Winchester, Kentucky over two days. We looked at cows and learned how to evaluate body confirmation, including: udder size and shape, hair swirls in the coat, movement and shape of legs and feet, and proportions of body parts to each other. It was fascinating. The goal is to select cows that produce plentiful amounts of butterfat. Calves that receive such carry more intramuscular fat when they mature, producing tender, flavorful meat and fertile bulls and cows.

One indicator of high butterfat in the cow is an active thymus gland. The gland is located between the chest and neck, and when fully functioning, shows up as a shining swirl of hair on the lower neck, such as on the cow in the center above. One would accordingly predict she will raise a healthy calf, who will in turn give plentiful butterfat as a cow, or provide excellent beef as a steer, or excel with high fertility as a bull.

This cow shows good confirmation because, among other characteristics, her flank girth (circumference just in front of hind legs) is greater than her heart girth (circumference just behind front legs). The heart girth should also measure longer than the length of backbone. These are indicators of high butterfat and high fertility. In contrast, a fertile bull presents a larger heart girth than flank girth.

We also witnessed how to carve a side of beef, beautifully demonstrated by a professor from the University of Kentucky. On the right side of the carcass lies an inside skirt steak. Afterall, the point of these beeves is to provide nutrient-dense food, so here it is, in raw and brilliant form.

At the conclusion of the second day of the meeting, an auction of purebred Red Devons was held. The stock were provided by various producers attending the meeting. Since we typically buy crossbred cows, which are cheaper than purebreds, to breed to Red Devon bulls, we were looking forward to another learning experience as spectators, when we attended the auction. We did not register and therefore did not have a number with which to participate in the bidding, fortunately.

Two or three animals sold at what seemed like depressed prices, but given there were quite a few cows to go, prices would certainly average out, we assumed. When the fourth cow came in and the auctioneer began to wax about the value of the animal, I detected motion to my immediate left. But being cool, I didn't break pose, even though the auctioneer was suddenly barking toward me and heads were turning to look at us. But we couldn't be bidding because we hadn't filled out the paperwork and didn't have a number. Not an issue, I told myself.

Yet I sensed the motion again and the barking my way recommenced. I began to suspect something was slightly awry or even more than slightly. Was I perceiving correctly that the auctioneer and my wife were in dialogue, along with some guy across the room? I tried motioning to her with thumbs at my waist, so no one would see, to cease and desist, but that turned out to be futile, as she was not paying attention to my thumbs. So, I summoned courage to turn my head ever so slightly to the left to catch her eye, only to witness Susan's hand up high, elbow moving rhythmically, and eyes locked on the auctioneer, as he sang his song, bobbing between her and the other side of the room.

This induced me to break into a sweat and start adjusting my hat, which helped, trying to look like we had this all planned, because the eyes of the whole house were suddenly upon us. I tried again to give her hand signals down low, but that proved as useless as the first time. I was trying to be stoic and impassive, like the other men in the room, but moving into slight panic, my head began to twitch and the whites of my eyes began to show. Finally, finally, the torture stopped, when the sonorous auctioneer waved a large arm our way and exclaimed with a flourish, "Sold!".

What a relief. I pulled on the visor of my hat, looked down to study thumbs, and just tried to breathe my way out of this. I faintly heard the auctioneer ask for her bidding number, gaining a rush of hope we'd be thrown out of the barn. She replied in full voice and without shame that she didn't have one, and within 30 seconds registration papers and a bidding number landed in her lap. It's okay, I whispered to myself, it's just one cow.

The next cow came into the ring and was met with the same sluggish enthusiasm. Uh oh, motion to my left, barking suddenly coming our way, then away, and then back again. People were beginning to smile at us, and cool as I knew myself to be, it was clear to the crowd that Susan's right-hand man was struggling for equilibrium, to the amusement of all. So, finally realizing I was vastly outnumbered, I gained sense enough to submit, and with a ready smile then decided to enjoy the rest of the journey.

So, we ended up with two beautiful purebred cows who are to calve within the next 30 days, at a very reasonable price. And I felt proud of my partner, who doesn't follow rules. After returning from settling paperwork on our new bovines, I observed a group of women and men surrounding Susan, congratulating her for bringing brazen vitality to the auction ring.

As our DiGiorno freezer has proved inadequate for continuous use, and as our inventory builds, we have invested in a second freezer that is 10 feet by 8 feet by 20 feet. By the end of the month, we will have another 150 whole chickens to store, for selling

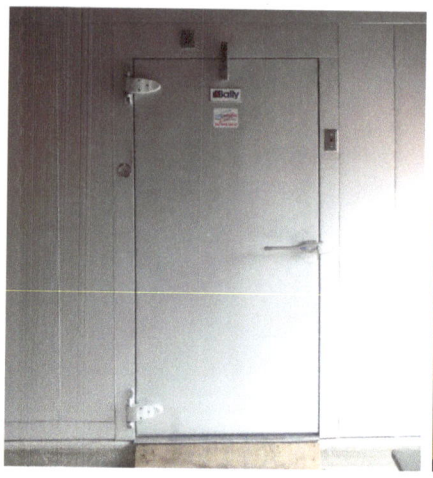 

throughout the winter, and this second freezer will be necessary. This is not a small investment, and we make it to serve you better.

My brother took this picture recently of noble dairy cows in Switzerland, returning to their valley on an ancient, narrow, and precipitous trail, after a summer of alpine grazing. Note the immense bells hanging from their necks.

Their journey reminds me in part of ours—the view is magnificent, the air is thin, the way is steep, dangerous cliffs await, and a village of wonderful people lie in support, enabling the journey to unfold.

Beef and lamb bones on their way to becoming grassfed stock for Susan's Soulful Chili.

Shimmering Spires
October 30, 2014

As these spires shimmer their way to the heavens, we bow before them.

This is a place of worship on our farm. Services here are conducted in strange cadence by invisible figures. Services are both endless and brief, structured and formless, liturgical and irreverent, comprehensible and unfathomable. The cathedral offers windows that transport one to the heavens. In this sphere, awe, humility, and peace blend to inspire the beholden. This corner of our land helps us raise our food and bring it to you and it helps you find us. Through places like this, we are united.

The same goes for Music Hall, the Museum Center, the Zoo, and sanctuaries of nature. They need to be preserved to nurture connections within the larger community.

Tuesday evening, we received eight pregnant cows and two bulls from Virginia to augment our herd. The bulls are registered 100% Red Devon and the cows half Red Devon. Hopefully we can now "close" the herd and grow the balance from within. This represents considerable investment, which we make to provide a steady supply of grassfed beef for customers.

Our previous bulls were underperforming, and we have learned that average bulls generate very poor returns. We look forward to seeing how these boys do their job and will keep you posted.

Last week we processed 100 meat chickens at Pheryl Zimmerman's farm. It is hard to imagine a cleaner or better process than this family executes. They line up with their roles and choreography, employing sure movement, sharp knives, stainless steel tables, cold well water, and caring attention. We receive the cleaned birds at their farm in the chill tank, lift them out and into plastic bags, pack them into coolers on the back of our truck, and transport them two miles to our farm. There we vacuum seal the bag, weigh the bird, and put it into our large freezer. Probably no more than 3 to 4 hours elapse between harvesting and freezing, with very few hands touching the bird. I don't think processing of chicken gets any better.

And the results are phenomenal—we have never tasted better chicken. I resist employing hyperbole, but this chicken deserves it. This kind of food cannot be procured at a grocery store, but only through the handful of farms in the region who take the trouble to move birds to fresh grass every day while feeding non-gmo grain to supplement. On-farm processing makes a considerable impact as well on quality of meat.

There are many ways to cook a five pound chicken, but we have found good results with: roasting uncovered at 450 degrees for 15 minutes and then 350 degrees for 45 minutes.

It couldn't be more simple.

Digging Deep

November 28, 2014

This was the ninth hole we dug, to solve the problem.

The well-pump at the dairy wouldn't stop running, so we assumed a leak had sprung in the underground waterline somewhere. Even with 18,000 feet of waterline to trace, a leak usually announces itself through a large wet spot in the ground. We have always found leaks in the past, but this one was proving totally elusive.

Landis and I looked and looked and looked, but found no wet spot, which generated increasing anxiety. A crisis I dread is having the water system go down in the winter, with 80 heavily lactating dairy cows standing by, needing to drink and be milked. Such crisis has to be solved immediately, even if it is pouring rain or 5 degrees below zero. A keen memory of misery is being four feet underground in a muddy hole in the cold rain repairing a waterline, with numb feet and hands. I could see the same scenario unfolding in this situation and resolved to fix this nagging problem before the weather turned inclement.

So, we called in the backhoe, and started where we discerned the elusive problem might lie – at the intersection of a tile line and waterline. The waterline could be leaking and the tile line would be directing the water away to the creek, rendering detection from above impossible. We did identify one tile line on the farm that was running, while all others were dry. So, we started with that tile line, and went to the first of two intersections with the waterline, and dug a four foot by six foot hole with the backhoe, only to find the intersection dry as a bone. Not discouraged, we proceeded to the second intersection and dug another four by six foot hole, to find it slightly wet but from ground water rather than a good-size leak.

Feeling stumped, but not deterred, we advanced ten feet forward along the waterline and dug two more grave-like holes, which also proved to be dry. This was beginning to look grim. But I came up with an alternate brilliant insight, so we raced over to the most obvious prospect of all, only to find the deep soil there dry as tinder. As desperation began to rise, it seemed prudent simply to keep digging, so we dug four more holes, each the size of a large sofa, all with the same result. By this time, the smile on the face of the backhoe operator seemed to be getting bigger and bigger, while mine had disappeared entirely. Hitting saturation point, I called off the day, feeling frustrated and somewhat incompetent, wondering if in defeat one ought to admit defeat.

That evening, the specter of accidentally being buried in a muddy hole in January, trying to fix a waterline, motivated further insight. I combed through my twenty year memory-bank, and suddenly realized the plugged tile line we found on the second dig was the eighty-year-old version made of clay, not the black perforated one we installed 20 years ago, 10 feet beyond.

So, the next morning with high hopes, the backhoe operator and I returned to the second hole, moved ten feet back along the waterline, and dug deep to find the elusive problem—water spurting out of a crack in the PVC directly into the tile line! What a relief!

The leak was caused by a trencher that installed the black plastic tile line above the waterline twenty years ago. The trencher left a slight dent in the PVC when crossing it 20 years ago, and that dent eventually turned into a leak. We cut out the piece of dented PVC, pictured lying on the bank to the right.

As mentioned, I felt tremendous relief at solving the problem. I also felt foolish for being so close on the second dig, but not realizing it was the black tile line we needed to find. But it had been twenty years since looking underground in that location, so we tore up more pasture digging seven more holes, and spent more precious money, time, and energy chasing our tail. In the end, we solved the problem; we put it to bed, and can now go on to more pressing matters, like marketing.

Why belabor this prolonged story and call on your precious time to consider it? Two reasons: first, it was a day in our life on the farm. Farm life offers much romance, but also offers many acute challenges. Being willing to climb into mud holes six feet deep is part of what it takes to deliver good food to you.

Second, this is a story to which we all can relate—it is the story of life. We have to persist foolishly in the direction of our dreams, even when it involves making mistakes before others, being inefficient, and exhausting ourselves. We have to be willing to believe in our unique vision at all times, such that the mind never stops wondering and never stops searching for better solutions. If we are willing to offer blood, sweat, and tears, the soul will grow and the vision will become evermore clear. This is true for the journey of bringing up children, loving elderly parents, advancing through demands of marriage, holding a job, starting a business, cooking a meal, or writing a poem. If we are willing to dig deeply to repair leaks, sources of life will flow.

This six pound, pastured chicken and evolving apple pie provided for Susan and me this Thanksgiving. We were solo for the first time ever and savored the tranquility.

In this season of gratitude, we express heartfelt appreciation to our customers for your support. We are most thankful you are our partners.

From the trenches.

Artisanal Logging
January 8, 2015

Drover, Arborist, Sawyer and Equines artfully harvest timber from the forest.

It has been 30 years since we last harvested timber from our woods, despite monthly inquiries from loggers. The venture three decades ago left us with scarred trees, deep ruts, and considerable apprehension.

We have pondered since how to manage our forest. We have placed a conservation easement on 100 acres of woods, to shelter headwaters of our wetlands, so those trees will never be harvested by man again. The rest of the forest is regrowth from 100 years ago that has been too frequently harvested. The forest has been highly manipulated over the years, and begs the question of whether to manage it to: recreate original conditions, leave as is, replant or replace increasing numbers of dying species, or simply harvest what is there every 20 years as a cash crop. We haven't been sure, and have erred on the side of doing less than more.

Trees take so long to grow, and contain such wealth of knowledge and beauty, that it feels sacrilegious to take them down. Especially when a tree that is 70 years old can be felled in 10 minutes with a chainsaw.

The proportion between time for growth and time for felling is extreme, even absurd. A tree is one of nature's greatest creations, and it is painful to lose even one to recklessness.

But we need to learn and resolve the dilemma between active and passive management of the forest. So, when a young Mennonite man approached me this fall about using horses to harvest some timber this winter, I agreed to a trial run of one load of logs. We are stepping into this to observe.

Notice the path through the woods the horses create. It hardly disturbs the underbrush, nor do the horses scar trees as they pass by. The operation is quiet, and does not exude dissonance with diesel fuel belching from sylvan depths. Horse-drawn logging feels measured and reasonable, somewhat akin to slow cooking—it almost always turns out right.

Our bovine friends constitute yearlings and two year olds calmly standing around, despite zero degree weather. We allocate a new strip of fescue to them every day, and they are thriving. In another week, we will start feeding them hay, as their allocation of grass will be consumed.

We still have another 40 days of grazing in front of the cow herd. Despite this, we started feeding haylage (fermented wet hay) to them today, so we wouldn't have to move them down the road and set up electric wires for the next pasture in this extreme cold. Once the Arctic front passes, we will resume grazing. It is a challenge to work outside in this weather for very long.

 We grilled a sirloin steak the other night along with a few pieces of tenderloin, creating sumptuous leftovers for appetites hungry from the cold.

Little House on the Dairy

February 7, 2015

Decent housing is central to decent living. It comprises one of the three basic needs, the other two being food and love. During the last half of 2014, we renovated this 100-year-old farmhouse, so Landis Weaver could move into it in January. The house was transformed from being dark and tired to upright and happy. For the past three years, Landis has been working from dawn to dusk to build the organic dairy on our farm,

and is thus deserving. He is also courting a young lady from a Mennonite community in Maryland, and is thus awaiting. So, he and we are prepared! Good housing attracts good people, as does good food and good loving.

We fed cows haylage for several weeks in January until we removed bulls last Tuesday, and then sent them down the road for another month of grazing. It doesn't look like there is much to graze on the brown hillsides, but green fescue lies beneath, which they covet.

Below is our last chicken, rendered into leftovers with rice, tomatoes, and eggs. Susan refers to this dish as "farmhouse chicken."

Beauty of Fat

February 23, 2015

Fat has many virtues, one of which is to provide warmth on cold days.

Fat on the backs of these cows insulates them from inclement weather. It offers a form of protection, enabling them to reside in fresh air and full sun at all times, despite freezing temperatures. We are selecting cows who store enough back-fat to make it through the winter without pampering, while nursing calves at the same time. Fat provides shelter.

Note how big the two calves are in the foreground. They are four months old and are obviously receiving plenty of butterfat from their dams to withstand these temperatures without loss of well-being. Fat on the back of the cow enables her to provide essential calories to her calf.

Fat also provides essential taste. I recall thirty years ago my mother was told to go on a low-fat diet to reduce risk of heart attack. She loved dairy products and meat fats, but nonetheless, she faithfully obliged, bowing to prevailing medical wisdom. Suddenly margarine showed up in the kitchen, and I remember tasting it on toast, and by the third bite it had given me a headache. I never touched it again. Margarine is a

polyunsaturated fat made from soybean oil that is a total stranger to our bodies. In contrast, our bodies recognize monosaturated fats from animals, as we have been consuming them since man slayed his first wild beast.

Prevailing medical wisdom of the past fifty years has sought acceptable explanations for rising incidence of heart attacks and strokes. It chose animal fats as the source of evil, and launched a well-orchestrated and charismatic campaign against it. Data provided was casually selected and not replicable, but in the absence of resistance and a credible alternative, the campaign won the day, making its way to the USDA pyramid of healthy foods, where meat and animal fat were essentially dismissed.

What wasn't being acknowledged during this time was the rapid growth of processed foods introduced into the dietary culture of Americans. These included refined breads, cakes, crackers, chips, and cereals, as well as the growing market for soft drinks. In conjunction, agricultural economics began promoting confinement-feeding of ruminant and non-ruminant animals, to consume ever-swelling supplies of corn grown in the Midwest. So, all of a sudden, grain-fed animal fats, high-fructose drinks, and refined cereal products swamped the American diet, enriching food processors, while starving the populace of traditional basic calories. The result has been obesity, heart attacks, and strokes way above historical norms.

In concert, American fashion adopted during this era a bias toward a skinny body type, particularly for women, severely injuring the self-image of our daughters. This conveyed they should be other than they naturally were, causing generations of pain for women and families. If Ralph Lauren wanted to make an enduring rather than fleeting contribution, he would change the shape of the women walking down his runways.

Mankind has been eating grassfed fats for two million years, while eating grain-fed fats for about fifty to sixty years. The former is high in

Omega-3 fatty acids, as are wild fish, and the latter high in Omega-6 fatty acids. We readily digest the former and have trouble with the latter.

Grassfed animal fats are part of our DNA. Our bodies recognize them and depend upon them to thrive. The trick is finding animals today that will provide this fat for us, while eating only grass. The quality of grass has to be high and the genetics of the livestock have to respond, to generate intra-muscular fat, which gives meat flavor and calories. Many breeds of bovines have been altered for grain-feeding so that good grazing animals are now hard to find. But we are slowly finding and raising them.

A well-researched book on the subject of the American diet is: *The Big FAT Surprise – Why Butter, Meat & Cheese Belong in a Healthy Diet*, by Nina Teicholz. It is a bit dense, but she has clearly done her homework to reveal the damaging fraud wrought upon our dietary culture by the medical, academic, governmental, and agribusiness communities since about 1960.

The picture on the left is of a recent rendering of gelatinous bone broth. The layer of fat at the top is being skimmed off and saved for cooking. (Almost all of our cooking is now done with animal tallow rather than vegetable oils.) On the right is curried lamb, made with stew meat from the fatty shoulder, underlain by Basmati rice.

As a side note, I entered our walk-in freezer this afternoon, set at 10 degrees Fahrenheit, to collect ground beef and lamb for chili, and found it warmer in the freezer than it was outside! That was a new experience.

In honor of mothers and daughters, may we celebrate the ancient custom of savoring fat from grassfed animals.

Pulling Posts
March 19, 2015

Pulling posts to make way for the new is exhausting work.

We had intended to replace this aged fence this winter, but frozen ground prevented the work. With the recent thaw, however, we spent Tuesday pulling 60–70 posts, leaving me plenty fatigued. Despite essential help from equipment and partner, Brendan, each post had to be manually handled several times. They are deceptively heavy, with some weighing

in excess of 100 lbs. The in-ground portion is usually saturated with water, providing weighty reluctance to clearing the old in favor of new.

So it is with life. Old barriers to successful living usually stand in plain view, and, at some point, it becomes worth the great effort to remove them. Withered and weathered structures are often perched on deceptively heavy foundations, so their removal is not easy. That is not to say all which is withered and weathered should be removed. On the contrary, some offers such beauty and wisdom, they warrant utmost sanctuary. But it is our responsibility to sculpt our own landscape and remove obstacles that impede us, whether on the farm, in the home, or in the mind.

We will replace this 70-year-old, woven-wire, locust-post fence with a lighter one, consisting of one strand of high-tensile wire supported by 1/2 inch fiberglass posts. It will be installed without equipment and will be enforced by electric current. It much better meets the description of the landscape we are developing to support us in years ahead.

Recent flooding left these culverts eroded and stranded. We are repairing the road, so horse drawn logging may proceed in nearby woods. Creating and repairing landscape never stops.

With all the mud last week, we found ourselves tearing up pastures as we delivered hay to livestock. Between rainstorms, however, we were able to deliver a week's worth of feed in advance, now only having to move electric wires to allocate hay.

After a day of sculpting landscape, a man is often hungry and tired. When he is "baching" it, because the good wife is serving other causes, winning solutions to complex problems arise. Tuesday night's dinner: one pound of lamb ribs, broiled 6 minutes per side, one pound of beef liver, sautéed 30 seconds per side, 2015 vintage well-water, and a homemade chocolate chip cookie for dessert. Breakfast the next morning featured beef kidney, with a touch of lemon, salt, and pepper. Totally satiated and beautifully satisfied!

One virtue of grassfed meats is they require so little adornment. They don't need to be buried in spices to be exquisite. Simple works beautifully for them.

In tribute to removing obstacles from the landscape.

Land Stewardship
May 21, 2015

Land is primal. It is a source of love and misery. Wars have been fought over land since aggression first arose. Land harbors minerals and organic matter that generate wealth on one hand and destruction on the other. Civilizations have risen and fallen upon the stewardship of their soils. Above all, land produces food and water, and thus is central to the well-being of civilizations. Land mismanaged creates misery; land well-managed fills hearts and pockets.

As one drives through the Midwest at this time of the year, the demise of land and culture is there to witness. Land that is meant to be green with budding plant life is brown due to application of Monsanto's Round-up. All plant life is killed to prepare the soil for a monoculture of corn or soybeans, rendering these fertile fields less biodiverse than the starkest desert. For six months these rich fields grow one species of plant and for the balance of the year they grow nothing. The soil becomes sterilized, and the outcome is a simplistic source of food that is laden with carcinogenic toxins, causing ill-health throughout the food chain. What sense does this make? Any first-grader's intuition would advise against this folly, yet it rigidly persists.

This paradigm of management is an insult to nature and to civilization. These Midwestern fields, now fallen to industrial agriculture, were once vast prairies, supporting root systems six feet deep and complex plant life that fed large herds of buffalo and attending species year round. If we do not reclaim these rich lands with biodiverse cultural practices, they will fail and we will fall.

Fortunately, there is reason for hope, as small pockets of land managers are advancing by returning to the roots of land stewardship. These are grass managers and vegetable producers found at farmers markets. They are doing earnest work, redirecting the production of food to time-tested agricultural practices, that do not rely on chemicals, hormones, and antibiotics. They are building organic matter in soil to produce food for humans rather than mining it for export to feedlots.

On our farm, we have increased organic matter by 30% over the past five years and expect to increase the rate of gain significantly as we manage more livestock, now up to 110 one-thousand-pound animals. We have increased organic matter by tightly grouping herds and forcing them to trample half the grass they are offered. This in turn feeds microbes in the soil and builds organic matter, storing both water and carbon, with which to grow food for people. Organic matter is a cornerstone of civilization. And those who know how to build it are in a position of strength.

Liberty Bell
June 25, 2015

This lichen-covered Liberty Bell has served our family for sixty years, calling home workers, children, and parents from the beyond. Its post rotted this spring, and we have been contemplating where to place it next. During recent years, it has stood in the partial shade of a maple tree, enabling sculptures of lichen to cover its wetter side.

As our nation's anniversary of freedom arises next week, we each ring the bell which signifies our freedom to be who we are and live as we choose. What a precious gift to celebrate.

As you contemplate how to celebrate the Fourth of July, consider grilling a boneless leg-of-lamb. One of the virtues of grilling or roasting boneless legs is the uneven thickness of the cut produces meat cooked both rare and medium-rare, pleasing a range of palates. Below is a leg we grilled and smoked, which turned out great. Accompanying the legs were: baked apples, green beans, scalloped potatoes, ratatouille, cherry pie, Eduardo's cheese, and Blue Oven bread. Quite a feast, generated by Susan's restless and resourceful hand.

In tribute to revolutions which have brought: Liberty, Equality, and Fraternity... and good food! May they continue.

Magical Discourse
August 13, 2015

Magical Discourse, plus miles of manure, produce a scene like this.

I spent most of last week with children, siblings, in-laws, nieces, and nephews on the North Channel, near the Sault Ste. Marie, Ontario. The visit was brief, but dense with dialogue, both awkward and flowing, among individuals and as a group. One interchange consisted of a three-hour group discussion. During the long drive home, the recurring reflection was upon the unforeseeable value that percolates from these manifold conversations. I found myself ruminating on the magical powers that arise from such discourse.

What is the relevance of "discourse" to farming and food? Our primary mentor, Alan Savory (www.savoryinstitute.org) offers that economies centered on land and landscape which fail, do so, not because of adverse weather and prices, but because the people involved are not aligned in values, goals, and context.

Ascertaining this kind of alignment is the hardest work of all, far more than lifting stones. It requires skills not readily taught in schools—

listening, validating, articulating, imagining, and projecting, among others. But once the alignment begins to dawn, physical efforts upon the landscape become coordinated, advancing progress of the whole to benefit all. And then the power of a common vision is unleashed.

Deep discourse is the soil's best fertilizer, but miles of manure don't hurt! As witnessed below, we are composting manure and straw from Landis' calving barn to add organic matter to pastures. If we can both build organic matter and keep ourselves aligned toward goals, we are much more likely to outlast adverse weather and fickle prices.

Manure and straw to be turned and composted, before spreading.

One of my most important roles on our farm is to entertain our two aging border collies—Dally (age 14) and Nick (13). This typically consists of their riding in the back of my truck, as I navigate around the farm, while they cast seasoned opinion from their thrones. They rarely consent to performing low-level work, now that Bo has arrived.

Last Saturday afternoon, I had them in tow in the back of our Kawasaki Mule. We were in a pasture looking at cows, when "boys" from Wright-Patterson Air Force Base came roaring over the horizon, only several hundred feet above, scaring the heck out of all of us. They circled several times, dropped flares, and expended taxpayer-money engaging in other supposedly useful exercises. By the time they disappeared into the clouds, Dally had disappeared as well. The loud noise and intense vibration were more than she could bear.

We looked and looked, but did not find her until late Monday morning on the backside of the farm. She was totally disoriented, weak, full of burrs, and her hindquarters seemed paralyzed. We brought her to the house, gave her a bath, and tried to feel for the next step. She wouldn't eat or drink, and was having trouble breathing. It became clear she had come near to the end of her road. With spousal consent, I called the vet to make an appointment to put her down but the line was busy, so I tried again, and while doing so, began picturing the scene at the vet's—green smocks, sterile syringes of toxic fluids, and mumblings of insincere condolences. None of that felt right, so I hung up the telephone, and decided to do it myself—the hard way, the honorable way.

Brendan lent me his rifle, and I took Dally to the graveyard, where Matonka, the once massive guard dog, is buried. I laid her on the ground, and began petting and thanking her, while her dull eyes looked up and she fought for breath. I kissed her, found myself starting to cry, and kissed her again. After another thank you, I mustered resolve, fetched the loaded rifle, brought it to the appropriate place, and pulled the trigger. Within minutes, her living agony was over, and the body lay perfectly still.

I unloaded the shovel and spud bar from the truck, and then began to dig. It took an hour or more in the hard dirt, and while doing so, the three of us talked: her body, her spirit, and me. I could feel her spirit about six feet up, gently hovering and watching, as I dug and sweated and observed, in the heat of the day. Her patient body was so peaceful, lying nearby in the sun, in marked contrast to the agitation she had previously endured. This all felt true and just and profound, in its own small way.

By the time, I had completed the grave, placed her in it, and covered her up, I realized I had experienced the magic of discourse as never before.

Dally

This week we have also been extending our underground watering system by 1,000 feet. As our cattle herd grows, we are needing more water under more pressure in 60 acres than a one inch, above-ground line provides. So, we dug a trench for a two inch water line, at the end of which stands a new cement pad, that will hold a 300-gallon tub in the winter. For a grazing operation, investments in water systems, such as this, and fertility of fields, as described above, generate the best returns.

Landis' eight-foot-tall organic corn, planted with horses in mid-May. During my several road trips this summer to Ontario and back, through Ohio and Michigan, no corn looked as good as his on our very own farm. Landis will harvest this by the ear, with horses, and then grind it throughout the year to feed his dairy cows, as supplement to pasture.

Below are two ribeyes we enjoyed last week. Every time we have a steer or cow processed, we taste the ribeyes to make sure all is well with the meat. This was from a seven-year-old cow, and we are beginning to believe that age has its merits in steaks as well as wine and people. The ribeyes were some of the best we have had. The accompanying meal included: chard frittata, corn on the cob, and a peach-and-melon salad, dressed with honey, cilantro and lime vinaigrette.

This past Sunday afternoon, after the farmers market, we enjoyed perhaps the most exquisite culinary moment of the summer, thanks to our colleague, Tom Keegan, in part, and to Susan's talent, in addition. This is soft-shell crab, sautéed in Kerry Gold butter, and served with a wine sauce, including blueberries and peaches, atop a bed of Becky's micro-greens. The bursts of flavor were astoundingly subtle, rich, and smooth.

We enjoy our magical discourse with you.

Dancing Shadows

August 27, 2015

This noble Belgian in this verdant scene, speaks of many things, one of which is the dawning of evening's shadows. As the sun sets on clear summer evenings, shadows dance across the landscape, in celebration of the goodness of the day and the fleeting moment.

One of the many interesting aspects of Nature is the most complex activities and most biodiverse places are at the margin of field-and-woods and day-and-night. More activity arises among plants and animals at dawn and dusk than at noon and midnight. So, it is particularly inspiring to watch shadows dance across the evening's landscape, heralding busy encounters.

For some, however, shadows represent darkness and fear. This fear can be dissipated, by recognizing that shadows are temporary and, in fact, represent affirming movement toward renewal. If you find yourself feeling marginalized, in the world of dark and confusing shadows, remember that in Nature, there is no place more rich with life than at the margin. So, you are probably in a very good place, that will inevitably lead to the light of dawn.

Below is featured beef skirt steak, spinach souffle, tomatoes, beets, and the stalwart of summer, fresh corn on the cob. The skirt steak is rich with fat, like a ribeye, but with finer texture. It is truly delicious. The problem is there are only two skirt steaks per animal, so they are always scarce. This was a fun summer meal.

We processed pastured chickens this week, so have a fresh batch available for your larder. Because they are a seasonal product that cannot be grown in the winter, we are trying to raise and hold enough in inventory to carry us through the winter. If you have space to hold them in your freezer rather than ours, that would be helpful. It is an excellent food that cannot be found in supermarkets.

Remember, we are among dancing shadows.

Country Music

September 25, 2015

The spectrum of life on our farm is rich with: decay, death, birth, rest, regeneration, nurture, and bloom.

Turkey vultures sort through the compost pile, promoting decay. The lamb flock is in full bloom. A pasture that has just been heavily grazed stands in contrast to another that has been rested for 30 days. Then we witness newborn calves, who have arrived through the mighty birth canal, to begin their great journey, and be nurtured by great mothers.

All of this is going on at once, creating a powerful symphony of activity, each representing a different force within the spectrum of life. This spectrum is like an orchestra, producing resonate music that reverberates in our ears and hearts, the melodies to which we willingly dance.

The challenge is to be able to hear it all, while the radio in the truck serves as a distraction. Listening is the hardest work of all in life, but as we sharpen our skills thereof, we are increasingly motivated by the beauty of the music produced by our land.

It is reputed that each valley in Appalachia creates its own accented banjo music. I am beginning to understand how that can be. With land producing primal vibration, combined with specific initiatives brought to it, like: birthing, grazing, dying, decomposing, regenerating, resting, and harvesting, unique piercing melodies arise, that never seem to end. These are the melodies that stewards of land and animals are sustained by and love.

Minor-key notes issued by decaying trees in our wetlands combine with those from budding swamp marigolds to generate accompaniment to the greater farm, like cellos harmonizing with violins.

Nutritious, delicious food is beautiful music in itself. Below we listened to the combined tunes of: smoked boneless leg of lamb, lima beans, corn garnished with sage butter, and cucumber and tomato salad. Emmylou Harris would have been proud.

Listening to country songs, we honor you.

The Tribe
October 9, 2015

Calves gather in the shade, forming their own tribe.

It is the nature of animals and humans to organize themselves in groups. People gravitate to their own families, clans, and tribes, but also to clubs, fraternities, sororities, alumni groups, associations, and, of course, religions. It is how we establish our identity. It is why emblems and symbols thereof seem to endure and proliferate. All groups are by definition exclusive. But it is okay that everybody does not belong to the same group, for those who do not belong to one belong to another. The trick is knowing where one belongs when. One's natural tribe changes during the course of a lifetime, which is fortunate, but can be confusing to the unaware.

Differences in groups create diversity in society and in ecosystems. If all humans were members of the same clan, with the same name and outlook, we'd be living in a vulnerable monoculture, not much different than the unstable corn field, sprayed with Round-up and exposed to erosion. We, in fact, depend upon differentiation, upon the tribal nature of Nature, to create a great number of alliances within the larger whole. This diversity establishes resilience against inevitable and occasionally brutal forces of change.

Hogs are never far from each other, often lounging side by side.

Last week our wedding anniversary rose to the fore. Susan prepared: braised carrots in bone broth, ratatouille, rack of lamb, and a spinach souffle—the piece de resistance. The souffle was made with a light bechamel sauce, pastured egg yolks and whites, fresh spinach, and cheese. It was silken. Why go to a restaurant when one can eat like this? There were embarrassingly few leftovers by the end of the evening.

Honoring your tribe.

Strands of Life
October 15, 2015

*We are enumerating "Strands of Life"
on oak beams in our barn.*

We summon these forces from within on a daily basis, so it seems appropriate to name and celebrate them:

    Dignity of Labor

    Depth of Heart

    Strength of Mind

    Power of Soul

    Beauty of Spirit

These currents seem to constitute the living experience, whether on a farm or in a city. We need physical strength, we have to care with our hearts, our minds must be engaged, and our essence must both receive and generate. Out of these efforts arises beauty of spirit. If we want to live fully engaged, all components of ourselves are called upon, whether wiping a baby's bottom, cooking dinner, buying real estate, listening to our spouse, commuting to work, writing computer code, playing chess, making hay, caring for animals, or stewarding children and parents.

The distinct art work above is being executed by Sarah Prendergast, our permanent artist-in-residence, to whom Brendan has the privilege of being married. She has sharpened her calligraphy skills from days-of-old, and is painting these inscriptions on our beams. We are most grateful for her talent.

I am not sure why it feels fitting to celebrate these concepts and place them in view, probably because they exalt the struggle of daily living, which is where we need the most help.

As cooler seasons dawn, roasted chicken becomes evermore beckoning. Here we witness creamed sweet potatoes, baked apples, green beans, and the last of cherry tomatoes, along with two 3.5-pound chickens. One of the most exquisite savory experiences, in my view, is to take a spoon, after a pastured chicken has been roasted, and scrape the pan for fatty juices at the bottom. They are so intense and rich, like aged cognac, but a lot easier to swallow!

There are many ways to cook a five pound whole chicken, but one successful method is: 450 degrees for 15 minutes, followed by 350 degrees for 45 minutes.

In exultation of the strands of life.

Animal Impact
October 22, 2015

The impact of bovines upon the landscape is powerful.

Two hundred thousand pounds of animal weight was applied to this hilltop for half a day by grazing cattle. The result was relatively uniform grazing of half the plants in the pasture, benefiting cows, and trampling of the other half, benefiting soil. Trampling of plants provides mulch to the soil, as does straw on reseeded lawns. This mulch keeps the ground moist and cool, which feeds microbes in the top few inches of the soil profile. Active microbial life composts dead materials and manure, creating organic matter. Organic matter stores 8 pounds of water for every pound of organic matter, providing irrigation to plants during dry spells. So, these cattle catalyze powerful forces beneath their feet.

We have not been as successful at energizing pastures with sheep, though we move them very carefully across the landscape. This is probably because each foot supports 1/10th the weight of the foot of a cow. And it is feet grinding the plant into contact with the ground that feeds the soil. Sheep tend to scamper around plants, while cattle push them over.

But each has its place on the landscape, with scamperers tending to win during droughts.

The dark spot pictured was a patch of dense, unfriendly thistles about five feet tall. The cattle in the distance were bunched closely enough, due to Brendan's management as a "grazier", to consume some of the thistles and lay the rest upon the ground, turning them into submissive mulch. In previous years, we have had to mow that patch of thistles to keep it from going to seed. This year, due to higher cattle numbers, we employed bovines to do the job.

High impact by livestock upon pastures must be followed by long periods of rest, providing plenty of time for plants to recover from grazing. Both of the above pastures were last grazed in July, and we are about to graze them again in late October and early November, which equates to 100 days of rest. Doesn't that sound enviable?

Lamb shanks being browned and then 9 hours later they sit in an irresistible sauce, waiting to provide succulence to the fortunate. These are easy to cook: first brown in frying pan, and then braise for nine hours at 200 degrees in a covered pot, steeped in beef stock and/or wine and accented with herbs of choice. It is that simple, with the taste being wonderfully silken, complex, and rewarding.

With appreciation for your impact on us.

Colors of Redemption
October 30, 2015

As the occasionally brilliant, but often muted, colors of Midwestern Fall emerge, they seem to correlate with our aging process.

If spring represents the hope of our 20s and 30s,
and summer tells of the labor of our 40s and 50s,
then fall reflects the wisdom of one's 60s and 70s,
with winter promising peace in one's 80s and 90s.

Despite one's age, this pattern presents itself every 12 months. In both the fall of the year and the fall of life, one feels as if matters are becoming settled. Supportive patterns have been established and quiet celebration permeates the air, due to bounty being reaped. Great colors bedeck the harvest, and as we witness the beauty thereof, we recognize that all feels right with the world for a few moments in time. That magic is unique to this time of year. Thank goodness for the sense of redemption provided by the colors of: red, yellow, brown, and green...

Work on creating our kitchen continues. This room once stored coal for stoking a furnace, but since the disuse of coal, has laid in dusty abandon. We knocked out a dividing wall, installed generous windows and doors, applied PVC siding to walls, and are awaiting countertops and sinks. A 60-inch stove will stand against the stainless steel, with the dot beneath the hood being for a waterspout, with which to fill stock pots. The other end of the room will hold two sinks, a dishwasher, and a prep table. More progress is at hand.

Featured on the previous page is a lot of polenta and a little bit of pot roast. Pot roast is easy and delicious during these cooler months. We have found successful temperature and timing to be: 200 degrees for nine hours, along with beef stock, some wine, and herbs of choice. Brown the pot roast before slow cooking. It is a true comfort food.

This week we move to the indoor Farmers Market at Clark Montessori, 3030 Erie Avenue. As winter dawns, please don't forget us and other vendors you patronized throughout the summer. It is hard to maintain a viable business in which revenues drop off precipitously during winter months. We are investing significantly to provide high-quality food for customers every week of the year. If the nutritional benefits of farm-raised food are important to you, please continue to vote with your pocketbook at the farmers market, which often requires intentional acts. At the same time, we know we always have to earn your favor, which keeps us continually striving to do better. We will only succeed by working closely with you, and we hope the quality of your life will only improve by working closely with us.

With gratitude for your patronage and for the colors of redemption.

Ancient Roots
November 6, 2015

The World Health Organization recently issued a vague and irresponsible statement: "Consumption of red meat may cause cancer."

What does that mean?! What are the specifics? What kind of red meat? How was it raised? What was it fed? Where did it come from? How was it harvested? How was it cooked? Where did the samples for this study come from, industrial agriculture or artisanal? Who conducted the study and how was it financed? Was it subjected to peer review?

Is the W.H.O.'s statement any different than asserting that: breathing air or walking in the sun may cause cancer? Aren't statements like this a form of indefensible hyperbole?

Scientific study has yielded great new possibilities for life on earth. But it is a very young and incomplete discipline, despite high rigor applied by best minds. Science is better at determining what is not rather than what is. Understanding what is requires identifying all parts of the whole and explaining complete interaction among them to produce given outcomes. This is an impossible task in the biological and natural world, where events are moving too fast and participants are too numerous to be contained and defined. Scientific methods are reductionist by nature and cannot incorporate holistic dimensions of the problem being analyzed.

How do we live with this statement by the World Health Organization? How do we make sense of it, so we are not propelled by fear, nor totally dismissive, nor wandering in ignorance?

My approach is to consider the issue by looking to Mother Nature to see what she has to say. Mother Nature has been at work in her laboratory for billions of years and has been successfully feeding human beings for 2 million of them. Two million years! Modern science has been at work for 100 years, perhaps 150 on the outside. Which data do you trust more – that from 150 years of intermittent research by an elite number of disconnected scientists in western countries, who sleep at night, or that from 2 million years by Mother Nature produced every day, 24 hours a day, with constant interactive feedback from all over the planet? There is no comparison.

One conclusion evolved from Mother Nature's exacting research is grassfed meats nourish the human body and are part of the solution to living a long and productive life. This is demonstrated by many examples of octogenarians and centenarians in aboriginal cultures, who lived off of migrating grassfed beasts – buffalo, wildebeests, antelope, camels, yak, sheep, and goats...

Farms like ours and those at farmers markets are replicating Mother Nature's time-tested model. Livestock are fed grass, some exclusively and others in supplement, and animals are constantly in motion. This motion replaces chemicals and hormones employed in feedlots to maintain health. This motion also requires a level of daily management the industrial food system can never provide.

None of us at farmers markets wants food from the industrial food system, which is why we interact with each other. The scientific study cited by the World Health Organization has to be employing data derived from industrial food because there isn't enough data available to date on the artisanal option. So, the study is basically irrelevant to us.

But use your intuition. When you look at these pictures of grassfed animals, how do you feel? Don't we resonate with visions of savannas, because that is the ecosystem we originally came from, that originally provided most of our food? Aren't we drawn to grassfed animals because of ancient imprints? This is powerfully true. We do not need to concern ourselves with hyperbole derived by myopic science on tangential and unrelated topics.

We are returning to our roots by eating grassfed meats, roots that are 2 million years old. These are roots we can trust; they are sound; they are tested; they are ones I am willing to bet my life on. And you can too.

Mother Nature knows best.

Pastured thigh & leg of chicken, creamed sweet potatoes, baked apples, and asparagus – by Susan's deft hand.

In ancient rootedness.

Dignity of Labor
December 3, 2015

Baldwin Piano Company

Epic mosaic murals, once at Cincinnati Union Terminal, dignify the working man. These murals depict robust men executing trades of industry in clean clothes and full light, prepared for dirt and sweat, while enhancing quality of life in society. There is something magnificent about these murals and their message, which is that working with both body and mind is glorious. What young child would not want to emulate those men? I certainly did. What parent would not want their child to do so?

Perhaps too many...

As our culture submits to the wonder of electronics, the virtue of physical labor is dissipating. Properly deploying one's body – back, legs, shoulders, and hands – to tasks of importance is an art-form, developed through practice and example. Though contemporary culture currently values mind over body, the pendulum will swing back someday.

Are not some of our fondest memories working outside or in the kitchen with a parent or grandparent, learning rudimentary but essential skills? These memorable activities tend to be low-skilled at the outset, but lead to high performance down the road. They lay the foundation upon which excellence can rest someday.

How would a delicious meal be generated from a kitchen without: cleaning, ordering, chopping, slicing, dicing, counting, weighing, sweeping, arranging, and setting? Each of these tasks does not require great skill, but executed together and in proper sequence create the foundation for the skilled cooking and delicious meal that finally arrives at the table. Initial steps cannot be forsaken, as the essence of the meal starts with them. When we taste delicious food, we recall it commenced with the mundane.

The same is true on the farm. When we sell you grassfed foods, we can trace the quality back to repetitive daily tasks of: feeding, moving, repairing, building, monitoring, and amending. Each task does not require developed skill and can be delegated to the inexperienced. But all tasks are connected, and if one is not properly executed, it can quickly spill over to the next to create a conflagration. Careful execution of daily tasks is essential to integrity of the final product.

Caring about each step, no matter how physical or uninteresting, is so important. The men in the murals seem to care. Men and women on farms represented at farmers markets care immensely, and therein lies the honor. It is highly honorable to perform manual labor carefully and skillfully.

Berea College, in central Kentucky, cites *Dignity of Labor* as one of its eight Great Commitments or core values. Private and state universities would do well to emulate Berea, which outwardly professes that excellence starts with manual and mental labor and the willingness to perform dirty work. Isn't that a different approach to education?

There is another dimension to physical labor, which is that our bodies are made for it. One of the attractions of farming is that it employs both mind and body. Farmers don't have to go to the gym to come into balance. But for those who do go to the gym, the activity is honorable, in that doing so attends to the body, one of the pillars of well-being, along with mind, spirit, and soul.

Dueling pots of nurture—grassfed chili on the left and Bolognese Sauce on the right.

In humility before the dignity of labor, yours and ours.

Savory in Georgia
December 17, 2015

Allan Savory,
helping us create the life we seek.

Last weekend we spent 24 hours with Mr. Savory in Bluffton, Georgia, home to White Oak Pastures, one of the largest providers of grass-based foods in the country. It was an honor to listen to such a wise man articulate profound, but intuitively recognizable, concepts. It has been remarked that "wise man" may be a contradiction in terms, but not on this occasion. There are so few such minds at large, that it is worth paying heed to one, when opportunity presents, so we did.

An incomplete attempt to summarize his thesis follows. The first step to creating a successful life is to describe the "holistic context" in which one wants to live: the physical environment, quality of relationships desired, values to be exercised, and general production to be realized.

One then employs one of four "tools" to enhance the sacred context: technology, fire, rest, or livestock. In order to decide which of the four to employ in a given situation, seven "testing guidelines" are applied to the problem at hand. They filter and select the scenario that most benefits the context desired. And thus the best decision is made. This decision-making process is invaluable on the farm, when so many factors are rapidly at play, asking for both daily and strategic issues to be resolved.

It is that simple and that complicated. Savory's 400-page text: *Holistic Management – A Framework for Decision-Making* fleshes out the details. More can also be learned by going to: www.savoryinstitute.org.

It was particularly poignant to be engaging with Mr. Savory while the Paris Climate Talks were concluding. We discussed employing holistic decision-making to rejuvenate the man-made, ever-expanding deserts of the world. Specifically, we brainstormed about using the tool of livestock to revegetate 100 million acres of the Sahara Desert, thus sequestering large amounts of carbon, while producing high-quality food. The discussions were stimulating and provide hope for some of the complicated problems we face as a culture.

Greenhouse

Sheep

Egg-laying Sheds

Delivery Vehicle

White Oak Pastures is an impressive place. The Harris family defies local agricultural tradition of irrigated corn, peanuts, and rice, by continually investing in pasture-based production. And they produce it all at a large scale: beef, chicken, pork, goats, lamb, rabbit, ducks, guinea hens, geese, eggs, and vegetables. The only thing they are not doing is dairy. They have their own on-site processing plants for beef and chicken, processing 35 beeves and 1,000 chickens per day. They hire 120 employees, whom they feed daily at their on-site restaurant. Most of their meat is sold wholesale. They execute at high standards. Their operation is more volume-based than we strive for, but we were privileged to witness excellence in action, which is always in scarce supply.

Our first pork chops! We cooked them last night, and thought they were great – sweet, juicy, and flavorful. The fat was soft and delicious, but not excessive, and throughout the meat, not just on the back. The meat was redder than industrial pork. This from our own Duroc hogs. So, fortunately they passed the taste test, which Susan holds at a high standard.

Thus this Sunday, we will have pork available for you for the first time. We have on-hand: two rib roasts of about 5 pounds, 8 cured hams of about 9 pounds, cured bacon, tenderloin, loin chops, rib chops, arm roasts of about 3 pounds, sausage patties, sausage links, and ground pork.

In tribute to all that is savory.

Depth of Heart
January 7, 2016

Passion usually accompanies success.

It takes a lot of heart to live passionately, for passion can be an unruly companion. Mistakes are made and energy expended prematurely, but living with a fully engaged heart is self-renewing in many ways. It enables one to persist through the unreasonable, adverse, and unexpected. A full heart is the source of courage, which is called upon to achieve success in most ventures in life.

Passion is central to long journeys, like: marriage, child rearing, elder care, entrepreneuring, pursuing a unique vision, or enduring physical or mental illness. Long-distance runners certainly possess finesse, but even more they possess heart. How do they continually push themselves through grueling thresholds of pain? By reaching into the only bottomless well they have—an endless desire to move forward, coming from deep within.

Life on the farm requires depth of heart. We are always wrestling with adverse weather and systems coming undone. Once the water system is secured in one area, it leaks in another. Once the fence is hot for the hogs, it becomes cold for the sheep. Once the weather cooperates, it changes. Healthy animals injure themselves and some die, despite heroic efforts. Machinery is fickle. Marketing of products is a constant mystery, as consumer demand is unpredictable, resulting in way too much product on-hand at one time and way too little at others. Whether it is 105 degrees or minus 50 degrees, animals must be attended to; there is no forgiveness. The soil evolves at too slow a pace. The song in our heads is ever changing. Resources are always thin, all of one's money is called upon, days are long, weeks are short, fatigue is an issue, and vacation is more concept rather than reality.

It takes courage to step into all of this, to persist, not to give up, to have faith in the long run, to love it despite and because of the challenge. One's heart launches this journey and mysteriously expands to maintain momentum. That is the marvel—the ever-expanding heart, for it carries us forward like wings of the dove. It is one of the strands of life upon which we depend.

Our newest member of the farm is Coquie. She is a Great Pyrenees guard dog, who is 3 months old, and is being trained to bond with chickens. We have begun losing fowl to coyotes, despite electric fencing. So, now we add another layer of protection, as we do with sheep. Our other guard dogs, however, are not eligible, as they are not bonded to poultry, and would see them more as feed than friend. Her name is derived from Coque au Vin, as discerned by our resident chef.

Speaking of the resident chef, Susan has on-hand for you exalted, grassfed, Bolognese Sauce, as displayed on the left. It has been said that such is one of the Roman Empire's greatest contributions to western civilization! And indeed it is – so smooth and elegant, accompanied with rice, noodles, or lasagna.

Grassfed chili is on the right – a rich and deep meal in itself, dense with vitalizing ingredients of beans, tomatoes, garlic, and hominy.

With full heart.

Strength of Mind
January 28, 2016

Our schooling teaches us to fill our minds. And it is a good thing, because the devil lies in details. If we don't tend to details, we trip and fall. The mind is usually at the heart of good judgement, as it examines alternate scenarios by which to choose the way forward. It distills complex realities into digestible conclusions. It unravels central questions and highlights potential problems. The ability to analyze stems from a disciplined and trained mind, gleaned from formal and informal schooling. It is frequently under-employed and occasionally over-employed.

Understanding all the forces at work on a farm is a tremendous intellectual challenge. Thomas Jefferson mastered the challenge, but few match his intellect, either on or off the farm. The rest of us strive to assess the deep concepts behind so much we encounter on a daily basis: biology of livestock, chemistry of soil, physics of construction and of mechanics, meteorology of weather, economics of markets, science of marketing, technology of computers, logic of accounting, skill of communication, trajectory of spouse, and, last but not least, technique of cooking. This is a hatful of hard information to wrap minds around, keeping students of farming fully and humbly engaged. One cannot manage a farm without the will to learn as much of the science as possible, which certainly strengthens the mind.

Isn't this true of every walk of life? Committing the intellect to the venture at hand usually generates positive results. But one of the curious aspects of the intellect is how incomplete it is, how it generates better results when working in tandem with its partners – heart, body, soul, and spirit. These are the five strands of life that weave the web of successful living. They are the iron horsemen with whom we forge the way forward.

One can't help being taken by the intense mindfulness of the tree-trimmer. I observed him lower to the ground a towering, massive maple tree over the course of eight hours, limb by limb. He skillfully interacted with a roaring chainsaw and tumbling limbs throughout the day, each 12 inches from his face, while steadfastly moving from one perilous position to the next. His mind had to be fully engaged, at every minute, for him to succeed, so as not to succumb to danger lurking so close by.

Landis Weaver and brother, Caleb, recently demonstrated both strength of mind and of heart by erecting this barn, square-and-plumb, on the coldest day of the year.

While this process would not be considered intellectual, he certainly demonstrated relentless ability to concentrate and make good decisions. He demonstrated the strength of mind of a seasoned warrior—the kind of person one wants on one's team. This man did not learn these skills in a classroom, but on the job through keen and persistent observation. His body was no doubt sore at the end of the day for its athleticism, but it was his display of disciplined mindfulness that was outstanding.

Pork shoulder, baked apples, rice and spinach, and a soft-boiled pastured egg. It took us three attempts to discover it takes 11 hours at 200 degrees to cook pork shoulder successfully. And then it melts in the mouth!

Bacon chunks display hearty and luscious constitution. These flavored Coque au Vin several weekends ago.

From the treetops.

Power of Soul
February 25, 2016

The Soul is a forgiving, honest, and healing recipient of our journey. It is a cauldron that receives our glory and our woes. It is where our blood and tears are collected. It is a sponge that absorbs our mistakes. It is a library that catalogs them, so we may return to them when ready. It judges not our folly; it accepts all, and is the willing host of our journey. It incubates wisdom, by being a patient observer of our experience. It makes room for our chaos and appeases it over time. It is the source of our healing, as it retains and cherishes the personal organic-matter we have shed through challenge, to be recycled into action when called upon. The soul is authentic; it tells no story other than its own. A soulful person states her truth, whether it be one of dishonesty or virtue. Her persona is clear and recognizable. The soul seeks expression, and withers under repression. Our egos often obscure the soul, casting a shadow over the deep richness marinating, percolating, and brewing beneath our surface. But as we discover courage, by tapping into organic-matter of the soul, we become honest about ourselves. And as we reveal aspects of our soul to others, we discover that soulmates abound, making the journey of life less lonely.

Power of Soul is a strand of life at our farm. We have experienced so much failure and agony in our journey, that we could not have persevered without the regenerative nature of the soul. For instance, during our second year of managing livestock, we were breeding 600 ewes. They began dropping lambs in February instead of May — four months early, which reflected only on my poor management of rams. Sometimes one escapes the consequences of such mismanagement, but we did not. That particular February was marked by torrential rains, which brought the creek out of its banks to flood the pasture in which the ewes were held, marooning them on an island. I was able to reach them with tractor and hay but arrived every morning to dead lambs strewn over the saturated and frigid pasture. I picked their tiny bodies up one by one, feeling the agony of death with each, and delivered them to a hillside graveyard. It was a penetrating and anguishing experience—a form of penance perhaps. I felt so responsible and irresponsible.

Later that year, we recognized we had to do more about the problem of foot-rot, that had arrived with one of the groups of sheep we bought and was spreading throughout our flock. It is an insidious and unfortunately infectious disease. So, Susan and I spent two cold, long days in December trimming all four feet of each of 600 ewes. That was an exhausting process, physically and emotionally.

We learned from both of those extreme experiences, and I am happy to report we now lamb in May and can say with confidence we have eliminated foot rot from our flock! But we couldn't have arrived at that result without the ability to persevere, which called upon our souls. And the remaining ewes who survived those storms with us are now truly our soulmates!

Last, it takes a lot of soul to simply feed animals every day, which Brendan does so well. We have over 200 animals across six classes of livestock, and to look at them every day and deliver to them their feed and to care about them constantly is a soulful act in itself. As is the feeding and caring of children.

These examples are no different from anybody's journey, really, whether in town or country. Children who face confronting obstacles in the classroom, on the playground, with social groups, or on the athletic field are feeding and deepening their souls. The same is true of adults, dealing with the challenges of work, home, children, marriage, illness, and caring for the elderly. Fortunately, the cauldron of the soul is always ready to receive!

Our wetlands has always felt like a particularly soulful place.

A farmers market dinner: Grassroots pork loin chops, Walnut Ridge carrots, Kristy's kale, Nathan's apples, and Elmwood sweet potatoes. We grilled the 1¼-inch chops for 2 minutes on each side and then put them into a 400-degree oven for 10 minutes.

From our souls to yours.

Beauty of Spirit
March 4, 2016

Beauty of Spirit arises from our work. It is the culmination of our daily efforts; it is our distilled selves. Our labor, heart, mind, and soul meld together to create context for the spirit. When the context is rich enough, the spirit emerges, like a plant pushing forth from fertile soil. And in whatever shape it emerges, it is beautiful.

The spirit is a positive force, whether in the self or on the landscape. When soil is not fertile, not much spirit arises. When labor, heart, mind, and soul are not committed, the spirit feels thin and weak. When these strands of life are committed, a positive, though imperfect, spirit may arise into the air.

Toxic soils do not present a spirit. The industrial, monotonous cornfields of the Midwest are forbidding. They do not beckon, as they have been sown with chemicals to kill, so inorganic fertilizers may preside. Whereas, a garden, rich with black earth and hard work of many hands, emanates a wondrous spirit. The same is true for prairies and healthy pastures, pulsating with deep roots, animal impact, and human attention.

This is also true for people, who tend to their work and reserve their judgment.

So, the spirit arises from within, from our work with the soils of daily life. I concede this is theologically a suspect point of view, but it is the prevalent experience to which I can attest.

There are times when the forces of Nature are so strong that the spirit within quakes before the spirit without. On one such recent, fleeting moment, I observed for about five minutes the sun casting its light upon the hillside. It created a vast and intense web of parallel shadows, racing madly but effortlessly up the hill and through the forest, to converge somewhere, at an impossible but inevitable point, in spiritual combustion. Don't we want to follow those shadows and discover that point? Look at the tremendous energy escalating through that forest. The whole landscape was alive and electric, as if a fire were ablaze.

Witnessing this made my legs shake. What is such an experience? Perhaps opportunities like this happen all the time, but we only receive and perceive them when we are centered enough to do so, when our soil is momentarily fertile, and our spirit can rise to embrace the wonder. Therein lies its beauty.

Thank you for indulging me, over the past few months, as I presumed to reflect upon the "Five Strands of Life" – labor, heart, mind, soul, and spirit. What more can be said about them, that hasn't been ventured by sages of the past? Most certainly nothing. But I offer these thoughts simply as testimony to intimacy with our land and to intimacy with you, for you are always nearby, even when we are far away. I am also grateful to Sarah Prendergast for her elegant inscriptions upon the beams of the barn.

Landis' dairy cow served as a foster mother this winter to our Red Devon calf, whose mother had abandoned him at birth. The dairy cow lost her calf mid-term, but produced enough milk to provide for two households and to let this Red Devon calf double in weight. Landis and Brendan successfully grafted the calf to the cow, which is not easily done, providing a resourceful solution for cow, calf, and people.

Moroccan lamb hash and poached eggs—delicious for cool weather! In honor of the spirit.

Table Travel

March 19, 2016

What happens around and upon a dining table?

As we explored in April 2014, newspapers are read, homework is conducted, groceries are stacked, food is presented, nutrition is administered, meals are savored, stories are recounted, community is gathered, values are taught, emotions are shared, concepts are pondered, plans are made, babies are conceived, babies are born, rest is taken, love is offered.

The primary eating table in a house is no ordinary piece of furniture. It is the most potent place in a home, as more goes on around it than anywhere else. One often-employed, but frequently overlooked, dimension to such a table is it allows home-dwellers to travel, as if upon a magic carpet. It is truly remarkable how far one can travel by sitting at one's own table.

When I first met Susan, she was a single mother of four very active, young children, who had led a fairly compressed life geographically. Yet, curiously the children bore a keen sense of the international world. Where did this heightened awareness come from, wondered the stranger?

This family was certainly not jetting around the world to fancy resorts. After a number of meals on their home turf, amidst happy chaos, this incongruous mystery began to come clear.

Each meal at their table was prepared by a creative, resourceful, and willing mother. Each meal told a story of some sort, and each story became elaborate in its own way, stimulating imaginations, and transporting the family to new lands—Italy, France, Spain, Ireland, Germany, China, Japan, Vietnam. The names of the food, how it was cooked, and what the menu symbolized were discussed, rolled around on tongues, and ingested in mind and body. Story-time didn't start before bed that evening, but a few days previously when shopping for groceries (which they did in a pack). As a result, these wide-eyed, eager children laced their conversation with ongoing references to: *guacamole, scallopini, escargots, tapis, borscht, dim sum, tagine, curry, uni, shitake, schnitzel...*, by second nature. They developed an unusual international vocabulary and sensitivity without leaving home.

When such children learn vocabulary and eat the food of a particular culture, they develop empathy and respect for it. This bodes well for the future, when they may find themselves in positions of leadership to engender connection and harmony among peoples, cultures, and nations.

The same is true for adults. Who needs carbon-burning jet-planes and exhausting, expensive international trips, when a creative hand in the kitchen can transport one to another culture, to be savored at one's own table during a relaxing evening? During the same time required to forebear intrusive security searches at airports these days, one can employ Table Travel to venture to India and back for an exquisite meal. So, why not?!

Pot roast on a wintery night. Where did that take us—perhaps to Robin Hood's Sherwood Forest, buried deep within English countryside, where archers relieve the 1% of their purses and maidens are protected in sylvan glens.

May your table always have legs for travel.

A Day's Movement
March 26, 2016

Nature is always in motion, sometimes imperceptibly, but it is constantly moving forward. On our farm, we strive to mimic such movement, as livestock travel in groups, across the landscape, throughout the year. This motion creates a proximity to wildlife and wild meat that is reflected in our meat. As with animals, so do people seek to move across landscapes. The following is a pictorial account of this person's movement on the Vernal Equinox.

Reflection

Patience

Contentment

Freedom

Balance

Power

Exultation

The fat on the brisket was as smooth and delicious as cream.

The motion of the day centered my heart. But one doesn't need a farm in order to move; sitting in an armchair can produce the same results.

May movement always be with you.

April Action
April 8, 2016

Now that March Madness is over, we head into the activity of April.

During April, the stage is being set for the growing season ahead. Once warm weather settles in, grass growth explodes, and then the race is on. So, preparations are underway at the farm in anticipation.

We purposely don't do much plowing on our farm, so perennial roots may grow ever deeper to enrich soil, but Landis Weaver plants 15 acres of organic corn to supplement his grass-based dairy cows. He leases the eastern half of our farm and is an intimate member of our small community on Frost Road. We figure there is less environmental cost to his planting organic corn with horses on our property than his importing corn from 50 miles away that has been worked with tractors and diesel fuel. He teams up with one of his colleagues and they combine teams of horses to do the job. It is always an inspiration to watch these magnificent equines at work or rest.

When pastures are sodden and weather inclement, we increasingly bring cows into a holding area to weather the storm. The problem is the holding area eventually becomes concentrated with manure. We have accordingly recruited several truckloads of wood chips from the local sawmill to spread over the manure, in preparation for the next visit from bovines. We will sprinkle corn into the layers of chips, and then turn hogs into the area to mix the manure with chips. The result over time will be compost, which we will then spread on fields. Holding livestock in a lot is a little like limited plowing of pastures, in that we resist doing so, but feel in particular circumstances, that benefits outweigh costs.

 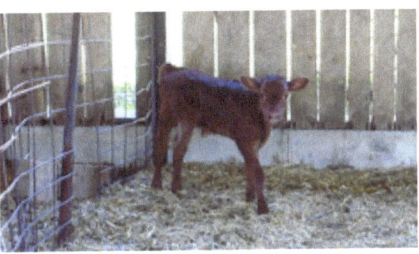

Landis' Jersey dairy calves were born during March and are now in the barn being hand-reared. We lent Landis two of our Red Devon bulls last summer to complete the last of his breeding. This recent bull calf on the right is half Red Devon and half Jersey. Red Devons fatten well on grass, and we are interested to see how this mix will perform in the milking herd. Someday we'd like to have a 100% grass dairy, with milk available to local customers. His milk is now sold to Horizon Organic Dairy and heads for parts unknown in a large tanker truck.

We are installing a crossing, across a spring-fed stream, to a wooded lot. The benefits of this for livestock are access to: shade during the heat of the summer, from the treeless side of the stream, and to flowing water, in the event our electric pumps go down.

Our first batch of Cornish-cross meat chickens in the brooder. Brendan fashioned this brooder out of unused space in the barn, and it has worked very well. The past several cold nights have been a concern, but heat lamps have seen the chicks through in good stead.

Delicious, soulful Hollandaise Sauce, made with our eggs. This was made with a blender, which is not quite as demanding as employing a double boiler. Hollandaise is a great complement for anything green—artichokes, asparagus, broccoli, and, for the dedicated, is rewarding on its own by the spoonful.

May the month be active for you.

Tragedy & Hope in Pike County
April 27, 2016

Intense tragedy struck Pike County this past Thursday night, 20 miles from our doorstep.

We have all heard some of the particulars, as the event made national news. This was an act so violent that it shakes the center of those nearby, either geographically or emotionally. To say the least, the event was extremely unsettling to anybody who cares about the flow of life.

What does the merciless squander of an entire family in a rural community reflect? Where could this phenomenon possibly come from? Is it not a symptom of an issue broad and deep in rural America? Has it not been creeping towards us for several generations? What is the source of this horrific unintended consequence? Are we wise enough to own the root of this tragedy? Do we dare ignore it and continue, yet again, with business as usual?

Wendell Berry anticipated this very kind of nightmare in his book, *The Unsettling of America, Culture & Agriculture*, written in 1986, 30 years ago. In it he questions the virtue of industrial agriculture, its efficiencies, and its specialists. He wonders how increased yields and fewer weeds will improve quality of life for a farming family. He questions how the removal of livestock from a farm and a landscape, in favor of long winters in Florida, will allow soil to rebuild and biological life, on which civilization depends, to thrive. He questions how one machine replacing the work of a dozen men will provide employment for twelve families. He wonders how the local school will stay open when twelve families leave for work in the city. He wonders what the effect on the community is when a capitalist from out of town, employing "economies of scale" rather than people, buys the local feed mill at bargain price. He questions why fertilizer now has to be bought from agri-businesses, rather than raised on the farm. He doubts the intention of seed companies, promising increased yields through hybridization and lawsuits against saving one's own seed. He can't imagine how animals that evolved to roam could be content confined to a concrete lot, being fed alien foods, the product of which humans are supposed to consume blindly. He can't understand how "bigger is better" and planting "fence row to fence row," eliminating rabbits, will help the community. He can't imagine how Californian vegetables at the local supermarket are better than those from one's own garden. He laments the industrialization of agriculture, and foresaw its undoing of rural communities.

Wendall Berry was indeed prophetic. The countryside is now distressed and very ill. Schools and shops have closed in small towns. Feedstuffs are grown for animals in distant lands, without a bite to eat for local families. Children are growing up malnourished and obesity is epidemic. Immense agricultural equipment stops traffic and financially overextends its owners. Economies-of-scale are thriving, creating operations of tens of thousands of acres, managed by a few men and endless amounts of machinery. Round-up is sprayed on fields every spring, saturating the soil and leaving toxic residue in crops and in animals fed those crops and in those grains and meats consumed by humans. Land grant universities are financed by chemical companies, skewing research. Local jobs are scarce, forcing expensive, lonely commutes to urban centers. Bright children depart, never to return. Main Streets are supported by itinerant tanning salons and thrift shops. Welfare and drugs have become the only reliable source of income with which to pay bills. Patches of marijuana plants and dens of methamphetamine become the last salvation, by default, not by choice. Sickness pours from the eyes of these afflicted.

Over the past fifty years, the efficiencies of industrial agriculture have all but annihilated the heart and soul of rural America, leaving its fabric exhausted and threadbare. The evidence is everywhere that this version of agrarian capitalism is destructive to agricultural communities. Most honest observers would concede that.

We need a new paradigm and new hope for the ailing countryside, still so exquisitely beautiful in accents of dogwoods and trillium. Fortunately, we have one and you are part of it.

Trillium and dogwoods give hope.

It lies in farms like ours and the Mancino's and the Eaton's and all others at farmers markets, whose farms lie deep in the country, serving as beacons of hope within regions of despair. Thirty years ago, our farm participated in the industrial model, and didn't even support one family full-time. Over the past 25 years, we have steadily converted it from producing food for animals to producing food for people. The land is now "certified organic" and we are rebuilding topsoil. The farm currently provides the majority of income for four families, a number anticipated to double in the next ten years.

What specifically can you do to mitigate the wanton deaths of a family of eight in rural Pike County this past week? First, you can care, holding the pain of this ongoing tragedy in your prayers and meditations, until it settles into perspective. This is no small task and may be more important than anything else. Second, you can provide support to farmers who are on the front lines in rural areas, so they may continue to serve as beacons of hope. You are already doing that, by buying their foods at farmers markets, and it makes a profound difference, about which you should feel good. Third, you can help find investors to secure land near farms like ours, so we can attract additional talent to train those who want to move out of despair and into hope, through production and marketing of grassfed foods.

It will take a large collective effort to revive rural America, but it can be done, one person and one organization at a time. Urban centers have a deep vested interest in the success of the countryside, so plentiful food, water, and wildlife will be secure into the future. Together we have already begun making a difference. Let us continue to summon courage to act, for rural America awaits us. It will not regain health without the hands and hearts of each of us upon it.

On rare occasions, we sample beef tenderloin for dinner. Filets were cooked for three minutes on each side in a hot skillet, in rendered beef fat, and accompanied by roasted potatoes, spinach, mushrooms, and béarnaise sauce. Mouth-wateringly good!

In the face of rural tragedy and in the partnership of hope with you.

Power of Water
May 27, 2016

Water is one of two essential resources.

On our farm, we are constantly managing water, so it may provide life for our animals, people, microbes, reptiles, amphibians, grasses, and trees. All of these combine to create a resilient ecosystem, in which to raise grassfed foods for delivery to you.

We must supply over 500 head of livestock and 3 families with water every day of the year. Nothing commands our attention more than an interrupted supply of water. Three different wells and one spring work to satisfy demand, all driven by electric pumps. When electricity goes out, water stops flowing, and heads jerk in response. Pumps also malfunction and springs and wells expire. In the past three years, one ancient spring and one newly-drilled well have gone dry, keeping us guessing, as to why water is not flowing.

Running out of water puts a strain on marital relations, especially when it happens in the middle of a shower or when the in-laws are in-residence. One Easter weekend, Susan's mother was visiting, and the spring feeding the house decided to stop providing. Fortunately, she grew up in the Depression, and adapted gamely to "doing without." Shortly thereafter, we drilled a new well, which provides bountiful supply, but carries silt that needs to be filtered.

What is worse is when 100 dairy cows need to be milked and the water supply ceases to function. One or two women pointing their finger at one is bad enough, but if there are 100, one is in real trouble! When supply of water for the dairy ceases, we have to diagnose where the problem lies, among a maze of underground pipes and electric wires. That takes some thinking, but we have always resolved the quandaries thus far.

Dairy cows drink a lot of water—up to 20 gallons per head per day. Milk pipelines and the dairy holding yard are washed as well after every milking, commanding total use of about 3,000–4,000 gallons per day for cows and dairy facility. Lactating beef cows require up to 10 gallons per head per day, and lactating ewes around 5 gallons. Our meat operation thus requires another 2,000 gallons per day. We use a lot of water in our kitchen, for cleaning pots and pans. Between three houses and an enlarged kitchen, we are probably using another 1,000 gallons per day. Total water consumption on our farm, therefore, approaches 7,000 gallons per day. All of this water is delivered by pumps at the bottom of wells and springs, through 32,000 feet of 2-inch buried PVC, and pushed by pressure tanks.

Smooth production of anything requires backup systems. We have finally decided to backup our wells and springs with county water, so this past month we installed another 1400 feet of water-line to connect to the county supply. The benefit of county water is it is chlorinated, which health inspectors require for commercial kitchens. It also provides its own pressure, is not reliant on the local grid, and supply is not limited. The liability is a monthly fee, the fact that it is chlorinated, and it runs endlessly if there is a leak. Neither livestock nor humans like it as well. But it does provide insurance.

We do not water livestock in streams except in emergency or in select times in the winter. We don't want cows loitering in and degrading this invaluable resource. We'd rather have the manure on the land.

In this trench, you will note two waterlines—a white two-inch PVC and a black one-inch polyethylene. The two-inch is bringing county water into the basement of the house. The one-inch is taking spring water out to new hydrants for livestock. We could have fed the new hydrants with county water, but health inspectors for commercial kitchens will not approve

of possible contamination from backflow from livestock tanks. We are using the same trench for two delivery systems—one for livestock and one for house and kitchen. This picture shows having to dig through a section of shale. We poured sand into that section to protect the pipe from sharp edges once the trench is filled.

Another aspect of managing

water is flooding. We typically experience flooding of our bottoms once a year and occasionally three or four times. We have to be vigilant livestock are not caught in floods and we need to plan for high ground to move them onto plus stored feed to offer, if necessary. The photos are of the same stream taken a little over a year apart. The black disc in the middle is a 300-gallon watering tub stranded in floodwaters.

We also have in our bottom lands a maze of 4-inch perforated tile lines,

buried three feet deep, on 50-foot centers, which have been installed over the past 50 years, for the purpose of draining residual water out of fields. This keeps our hydric soils, underlain with a seam of clay, from returning to wetlands. The picture shows an outlet for 30 acres of tile lines.

The big concern is drought. We have been fortunate not to have experienced the drought the West has known, but our day will come, especially with the pressing advent of climate change. We had a very dry summer in 2012, and want to be prepared when such weather returns. We would like to build several lakes with which to hydrate subsoil and serve select irrigation, but that is a capital-intensive solution. The best strategy by which to supply water during a drought is to build organic matter of soil.

Every 1% of organic matter holds more than 20,000 gallons of water per acre. This equates to about ¾ of an inch of rain. If organic matter is 5%, then the soil provides a reservoir of water equating to ¾ of an inch per week for 5 weeks. That is sufficient moisture to successfully come through most droughts.

We are increasing organic matter on our farm at a rate of about ⅓ percent per acre per year. Our average organic matter started at 2%, is now close to 4%, and we are aiming for 8. As stock numbers grow, the rate of increase should increase as well, as long as we employ high-density, low-duration principles of grazing.

The concluding comment about the power of water to sustain life is we are building a swimming pool. This is an investment in quality of life, which we have long considered, and increasingly value, as we age. The pool will be heated and have a cover to provide resistance-free, aerobic exercise seven or eight months a year. It is not for luxuriating next to, but for maintaining health and well-being within a busy schedule. Allan Savory encourages developing plans for items like pools, as he explains a landscape should support the life envisioned. And besides, a pool returns one to the amniotic state, where all has been well for several million years. So, it feels like a prudent decision.

Our meal on Derby Day, which is considered by my Kentucky wife to be the most important holiday of the year. In its honor, a repast appeared of: Kentucky fried chicken, corn pudding, asparagus, biscuits, and Derby Pie – with pecans, corn syrup, chocolate, and bourbon; enough richness to make your eyes roll backwards. What is not pictured is the Kentucky Lemonade, with which the two most exciting minutes in sports were watched.

The other essential resource is the sun... But today we express gratitude for the power of water to enhance life.

Swimming forward.

Ancient Wisdom
June 10, 2016

Ancient wisdom reveals itself.

Two years ago, we poured a cement slab next to our dairy, located on a rise of land beside a stream. Two months ago, I learned the operator of the Bobcat, when excavating the slab, found a "stone tomahawk" in the dirt, which he put into his truck and drove away with at the end of the day. Two weeks ago, I told his supervisor the tomahawk belonged to the property from which it came. Two days ago, the supervisor recovered it and returned the artifact. As he laid it into my hand, I felt dumbstruck to be holding something so ancient and powerful.

What does an implement like this mean? Where did it come from? Who made it? How was it fashioned? Why has it been resting on our property? How old is it? For what purpose was it used? Can its stories and mysteries be divined? Why did it come to our attention when it did?

How do we honor it? And what should we do with it? These questions circle.

This implement is made of granite, most likely coming from a northern latitude. It is either derived from Native Americans of the north, who were passing through, or from a local tribe, who had traded for the material. Given our proximity to Fort Hill, a site of prehistoric worship five miles away, one would think this implement stems from that culture, which no white man ever witnessed. How would they carve those smooth and nearly symmetrical features out of something so hard? What tool did they employ? Certainly not a hammer and metal chisel. Pictures of similar artifacts are dated at 5–6,000 years old. Imagine any tool of ours lasting that long. We are lucky if cell phones survive one year. Was this an implement of war, domesticity, or religion? Was it left behind by design or accident? What other implements accompany it, now protected beneath concrete? So many more questions arise than answers.

I distinctly feel this beautifully beveled, granitic, ancient, artifact of war and peace has a heart. One can almost hear it beating. It feels sacred and powerful in the hand, representing an immense span of time. Imagine all it knows and is willing to impart, if we only knew how to listen. Because it is impermeable, imagine how much more life it still anticipates.

We consider it a blessing to receive this ancient symbol into our midst. It is hard to understand how and why it happened, but here it is. It perhaps reflects immense knowledge lying within the land, awaiting recognition. It is confounding and stimulating how much there is to observe in any landscape, if one opens the eyes and is patient. Patience is the most difficult aspect of learning. Imagine how patient this carefully carved piece of granite has been over the millennium. And as patience is a characteristic of wisdom, it seems this artifact is a symbol of wisdom, offering infinite possibilities to us, as we learn how to receive them.

One clear sensation that arises from holding this five-inch-piece-of-history is humility. What are we, compared to it? How many hands, peoples, miles, and seasons has it known, compared to us? It is immense in experience, while we are small. Yet, we have intersected with it, for a brief moment in time, and are now partners in a journey ahead, until it outlasts us. What a deep and curious privilege.

Notice the stunning wooden board, racing with rich grain, beneath the stone artifact. The board is a cross-section from an olive tree in Italy. Olive trees are very long-lived, into several thousand years for some. This section of olive tree and this carved piece of granite seem to recognize each other. They produce harmony together, unavailable to contemporary matter. They now reside and preside in concert, momentarily on our dining-room table.

This is the first year we are not making hay. Instead, we are planting an annual, sorghum-sudan grass, into pasture to extend the grazing season into the winter, as an attempt to replace some hay. Seedbeds for annuals are typically prepared by either plowing or spraying. We are doing neither, but have applied "animal impact" in the winter to break up sod, grazed heavily just before planting in June to remove competing plant-matter, and mowed remaining plant growth just after seeding. The benefit is we are not reducing organic matter by tillage nor mixing herbicides into the food system. The drawback is the resident perennial grasses may grow faster than the newly planted annual ones and shade-out its growth. So, there is some risk. If this works, however, the approach would reduce reliance on hay, generating significant cost savings.

We are trying to eliminate the high financial and environmental cost of the picture on the left, by drilling annuals into pasture, as depicted on the right. If you look closely on the right, the lines of the planter or "drill" are evident, but seeds have not yet germinated. You can also see how the dense sod was weakened or opened-up this winter, to prepare for drilling this June. We will keep you posted on progress.

A recent repast of Bolognese Sauce with pasta, fresh asparagus salad, a mozzarella & tomato salad, and fresh strawberries from Elmwood, all presided over by an ancient beveled stone. The bolognese is remarkably clean, smooth, and elegant. Three cups of bolognese on one pound of pasta easily feeds four adults.

May we each be wise enough to receive wisdom when it surfaces, especially when it is 5,000 years old!

Power of Carbon
July 15, 2016

All living organisms are part of the carbon cycle.

This pasture was three feet tall just before grazing. Half was grazed and half trampled, with the latter creating a mulch that enhances the carbon cycle. The carbon cycle is the movement of carbon from the atmosphere to soil and oceans and then back to atmosphere. This cycle, when unimpeded, is at the heart of healthy ecosystems. It keeps our soils rich with biological activity and the food therefrom dense in nutrients. It has been refined by Mother Nature over the millennium. Humans are about 18% carbon, and all animal life is dependent on the movement of carbon through their systems. When this elegant cycle is interrupted, problems arise.

Conventional agriculture was the first to interfere with the carbon cycle. In about 3500 BC, the wooden plow was put to use. Tillage of soil releases carbon prematurely, increasing the load of carbon in the atmosphere while decreasing it in soil. This impact upon the carbon cycle accelerated over the succeeding 5,000 years. Tillage agriculture generated a lot of food in the meantime, but goes down in history as the first significant polluter of the environment.

Domestication of livestock also upset the natural carbon cycle. Confinement of cattle and sheep began around 4000 BC., degrading pastoral landscapes and releasing carbon from soils into the atmosphere. This practice of management reached epidemic proportions over the past 150 years, resulting in vast deserts throughout the planet.

The industrial revolution arose in the early 1800's, extending into the 21st century. During this time, fossil fuels were discovered and mined to generate heat, becoming the centerpiece for production of powerful and plentiful energy. The result was over the past 200 years, tremendous amounts of carbon, in forms of coal and oil, were harvested and burned, depositing abnormal amounts of carbon dioxide into the atmosphere. This set the stage for the climate change and global warming we are experiencing today.

So, tillage of soils, confinement of livestock, and burning of fossil fuels all have radically impacted the carbon cycle. Why do we care? Well, because too little carbon undermines the ability of Nature to function and too much carbon in the atmosphere will kill us! It is a potent element to be taken seriously.

And how is this relevant to your dinner table and our activities at Grassroots Farm & Foods? It is relevant because a herd of bovines, tightly bunched, delivering over 100,000 pounds of weight per acre, and kept in constant movement across a landscape, both rebuilds degraded soils and pulls excess carbon out of the air to store in roots of plants. It does this while providing nutrient-dense food as a by-product. This is an amazing feat that restores the carbon cycle and is unmatched by modern technology, providing triple benefit to society.

An acre of soil six inches deep weighs about 2 million pounds. For every 1% increase in organic matter, within that acre, carbon levels rise by 10,000 pounds. That carbon comes from the atmosphere and from the plants trampled onto the soil. This builds organic matter, which holds water for plants. Levels of organic matter on our farm have increased by one third over the past five years, now registering around 4%, with a goal of 10%. Our herd of bovines is the only tool we know to employ to achieve that goal.

It has been said that civilization owes its existence to its topsoil, from which its food comes. That remains as true today as ever.

When you buy food from Grassroots Farm & Foods, you are restoring the carbon cycle and building organic matter in Pike County, Ohio. That is a profound action, and you are our partners in this process, for which we are most grateful.

We enjoyed a restful week in Ontario, paddling a canoe, reading, and sleeping more than expected. One morning we set off to paddle around an island. We were greeted at the outset by a big bald eagle, alighting on the rocky shore 20 feet in front of our bow, with a fish in its claw! It was a powerful and dramatic sight. Neither of us had seen a bald eagle that closely, making that experience the highlight of the trip.

Lichen-laced Cliffs

Wild Irises

Yesterday we received a visit from the Ohio Department of Agriculture. The gracious woman said she was following up on an anonymous "complaint" that we are selling prepared foods from our kitchen, which had not been "inspected". I replied that is correct, and it is my understanding if one is only selling to the retail market, rather than wholesale, one doesn't need an "inspected" kitchen. I was advised to call the Division of Meats at ODA to confirm, which I did, and the understanding was confirmed.

So, what is going on here? Somebody is feeling threatened, wants us to cease doing business, and has reported us to the Ohio Department of Agriculture. I suppose we should take that as a compliment. At least they are noticing all the hard work. And they might further take note we are not going away. If the kitchen needs to be fixed in this way or that, we will do so. The critical factor is we work hard to produce food of high integrity and a growing number of customers are validating the effort. We and others have invested too much to turn back now. So, forward we go, into the competition!

A great summer meal: first ears of corn, fresh peas, fresh tomatoes, homemade French mayonnaise, and an irresistible Shortrib burger... It is hard to do any better than that.

May our carbon cycles flow unimpeded!

Goddess Energy
July 28, 2016

It is always prudent to have a goddess on one's team!

Goddesses inspire the unusual, provoke new directions, expand horizons, pick up their collaborators, see around corners, are tireless in the quest, provide endless support, are fertile with ideas, and can be totally entertaining. At times, they can be ever so aggravating, for they are usually right in concept and not always subtle in delivery. They often operate in their own dimension, creating protocol as they proceed, rather than adhering to the company line. Their inherent fearlessness can alternate with high caution, for reasons unexplained. They are thoroughbreds, who win races. If you have big arms, a goddess is a powerful ally on the team. You may wonder what you wished for at moments and you will need to hold onto your hat during the journey. They are not to be tamed, which is why they are goddesses!

The open spaces of a farm provide great places for goddess energy to be absorbed without crescendo. Wetlands, trees, grasses, forbs, living soil, and vital animals all provide natural companionship for the spirit of a goddess.

We have a number of goddesses at our farm. Susan now has a pool fit for a goddess, and she qualifies in many regards. Sarah is strong-minded and talented, bringing great energy to our landscape. She and Brendan are raising willful daughters, who already hold promise to make a difference in the world. Some of our animals present extra dimension to themselves, harkening to outer forces. Our lead guard dog, Abie, and our new one, Coquie, both are strong of spirit. Some of the lead cows are clearly exceptional in presence and performance. The ewes are other-worldly by nature, and seem imbued with spirituality. Hens are productive, elegant, and periodically demanding, so they qualify as well.

A farm with livestock is a place where the female spirit reigns supreme. The cycles of Nature are alive and active on such farms. On ours, it is a privilege to live among so much mother-force and goddess energy, for it is rich and productive, and is thus one of the primary sources of our well-being. So, bring on the goddesses! They are not always easy, but they are always great.

This hot and humid time of year stimulates growth in many areas, one of which is the herb garden. Susan's includes: mint, Thai basil, sweet basil, lemon balm, French sorrel, Italian parsley, lemon thyme, German thyme, rosemary, chives, savory, and Greek oregano. All of these are employed in her soulful products, such as: sliders, mayonnaise, bone broth, Bolognese Sauce, and chili. We enjoy American Sliders, which are all-beef and more mild than the Moroccan or Vietnamese.

Speaking of kitchens, we had a follow-up visit last week from the Food Safety Division of the Ohio Department of Agriculture. Even though the Meat Division has disavowed responsibility for us, Food Safety seems to want to adopt us. The lady "consulting" was a goddess herself, replete with winning tattoos, stemming from ear to foot. She was highly intelligent, fully sympathetic, and complimentary of the kitchen we have built. She did not issue mandates, but only suggestions. If we execute the following steps, we can gain certification as a Manufacturing Kitchen for retail sales:

1) create a Standard Sanitation Operating Procedure (SSOP) for our large stock pot, which doesn't fit in the dishwasher,

2) have the well water tested annually,

3) attach a hands-free adapter to our hand-washing sink,

4) provide paper towels rather than linen towels for drying hands,

5) provide a finished surface for wooden baskets holding implements,

6) label unmarked jars and bottles with a sharpie,

7) have our scale certified by the department of Weights & Measures,

8) list all sub-ingredients on labels.

We can implement these suggestions readily. That bald eagle we saw is providing guidance.

A distinguishing attribute of the goddess with whom I live is an insistence on celebrating life on a daily basis, preferably through good food, but also through other gestures. This can create a challenge for unimaginative souls accustomed to plodding forward into another day's work, but, I confess, it has proven liberating to try to meet the challenge. We recently celebrated daily living with a rack of lamb, on a Sunday

evening after the market. In accompaniment were: Blue Oven bread, mozzarella from Eduard and Sergio, baked apples from Dennis and Nate, ratatouille from Hazelfield, and egg frittata from our hens.

May the energy of our goddesses carry us forward!

Wedding Altar
August 26, 2016

We are building an altar for daughter Mary's wedding.

In early October, she, the intended, and a troop of friends and family will descend upon our farm. We are recruiting ancient forces from the Devonian Period, of 400 million years ago, to produce this altar, so she and he may successfully tie the knot. Now that is commitment!

Note the ripples on the sandstone, imparted by shallow seas that once covered this land. This stone is six feet long, six inches thick, and three feet wide. It was mined in a small quarry in the hillside across the road, by LEO, we assume, whose name is engraved on the stone. For the past 50 years, it has been resting atop and covering a shallow, hand-dug well on our property. Last week, we lifted the stone off with chains and front-end loader, filled the well with gravel, and carefully transported the stone and LEO to their new resting place, elevating them to renewed purpose.

If any of you feel inclined to reaffirm vows, we now have a place before which to do so and a stone in which to chisel your testimony. There is no need to rush, however, as the opportunity should be around for another 400 million years.

Sirloin steak and strip steaks, cooked rare. Grassfed steaks cooked beyond medium-rare result in tough eating. Heirloom tomatoes were exquisite. It is such a treat to experience their gentle, bursting, intense flavor, so unlike those found during the rest of the year.

In deference to ancient altars which await our vows.

Biodiversity & Humility

September 8, 2016

This cool-season pasture is rich with biodiversity, now heralding a newly-arrived, warm-season species, Big Bluestem.

This is the first year we have seen the tall plant in the middle of the picture – Big Bluestem – in this pasture. The pasture is typically dominated by fescue and clover, but there is also Ironweed, Cocklebur, and Johnsongrass. We seek as many diverse species of plants as possible in pastures, so livestock can choose among a buffet of calories as to what suits them best at that point in time. We also offer plants at different stages of maturity, for the same reason. Further, diversity of species is insurance against disease and adverse weather, with each species responding differently. It is often repeated that in Nature, diversity breeds stability. The more complex an ecosystem, the healthier it is. Thus, we welcome the arrival of Big Bluestem into our fescue pastures.

It arrived via bird or bovine from the neighboring pasture, where we planted a host of warm-season grasses five years ago – Switchgrass, Indiangrass, Big Bluestem, and Little Bluestem – which grow in temperatures above 80 degrees. Cool-season grasses go dormant in those conditions, so it is helpful to have warm-seasons to fill the void. They are hard to establish at the outset, but do spread once taken root.

Monocultures of corn and soybeans, which proliferate throughout the Midwest, will prove not to be sustainable. It is totally unnatural for one species of plant to grow in a field, that was once complex prairie, for 6 months and no species to grow in it for the other six months. Those conditions are worse than in a desert. Nature fights such paradigms of simplicity, and, in the end, will win the battle toward complexity and biodiversity, as witnessed by increased weeds now resistant to Round-up, upon which monocultures depend.

When you eat 100% grassfed beef or lamb, you are supporting biodiversity on farms. The more biodiversity farms generate, the more sustainable they will be.

Given that Nature is usually a fair model for society, it would seem that socio-diversity would be a stabilizing force, in the same way that biodiversity is. History has demonstrated that monocultures of people don't seem to persist for very long, in the same way that monocultures of plants do not.

Cows, preparing to calve, in their increasingly biodiverse pasture.

Life is humbling for most of us. Sometimes we have to be reminded of forces greater than ourselves that influence the journey. Some men want to overlook their age, and pretend they are as they always were. I experienced such a reminder several weeks ago. I drove 600 miles to Ontario and back in three days, and then jumped into a weekend of back-to-back 18-hour work days in 90-degree heat. Mother Nature thought this was rather foolish behavior for a 62-year-old man, and promptly sent the fool to a "time-out bed" for three days of vertigo and nausea.

As the world swirled during that interlude, I wondered whether enough wisdom would surface from within to listen to the message delivered. My capable uncle received the same tap-on-the-shoulder at about the same age. He rallied from the first setback to resume his former intrepid lifestyle, only to receive a subsequent setback, which sent him to bed for the rest of his life.

Mother Nature always has the last word, and it is humbling to be in her embrace. We just have to be quiet enough to listen. The noises of success and of love can obscure deep palpating messages that ask to be recognized.

On the third evening of being in the time-out-bed, I was feeling better enough to write about the Wedding Altar. In a slightly altered state, I misspelled Altar. It was embarrassing for this English major to make such a mistake, but life is full of mistakes. One can't be afraid of them, or one does not venture forth. It is better to try and fail, than never to try at all. Mistakes appear to make us look small and imperfect. But they really make us human. Our best humanity is in our imperfection, in our humility.

In an effort to listen to signals about overworking ourselves, Susan and I are privileged to welcome to this growing team a new employee and partner – Beth Gehres. Beth is an urban refugee from Cincinnati, like the rest of us, and spent 35 years working in nearly every aspect of the insurance industry, mostly with American Financial. The call-of-the-

wild brought her and husband, Bob, to a farmstead in Hillsboro fifteen years ago. Bob has been helping us part-time for the past four years with animals and landscape. Beth feels she has now completed her long mission with the insurance industry, and is eager to contribute her many talents, full-time, to issues of food, health, and land. Beth brings to us very welcome support in the kitchen, in the office, and at farmers markets. We are most fortunate to cross her path, as she is replete with: dignity in labor, depth of heart, strength of mind, power of soul, and beauty of spirit.

Susan's Rio-Grande Beef Barbacoa making its way onto cornmeal tacos. This is so darned good it is hard to know when to stop eating. The barbacoa is made from shoulder of beef. It sits in a dry-rub of spices overnight, is browned, and then braised for 12 hours in chicken stock and apple-cider vinegar and spices. It is subsequently pulled apart, discarding seams of gristle, bone, and undue fat, resulting in a 50% yield. The braising liquid is reduced and added back into the meat, to sit overnight in the refrigerator, for a final round of marinating. We then package and freeze the barbacoa into one-pound pouches. The meat is ready to be served, by bringing a pot of water to a boil, turning it off, and then placing the pouch in the hot water for 15 minutes. The preparation of this product is an involved process, resulting in a product with a tender, gentle, and complex flavor, that is addictive!

She creates Tar-Heel Pulled Pork by the same process, with the additional step of smoking the shoulder for four hours and marinating it in a barbecue sauce, that hails from the Smokey Mountains.

May we humbly advance toward a biodiverse future!

Wedding Celebration
October 13, 2016

We celebrated a wedding at the farm this past weekend.

Daughter Mary and her new husband, Leighton Zema, were wed this past Saturday at our farm. 180 people attended from all corners of the continent and one even from Dubai. Lodging was secured 40 minutes away in Chillicothe, and buses transported guests back and forth. We held a hog-roast on Friday night, a farm tour on Saturday morning, and the wedding ceremony, in dramatic shadows, Saturday afternoon. The ceremony was officiated by Mary's brother, my son, before the stone altar, bedecked with stunning flowers, and heralded by a flock of soaring, black, turkey vultures against a clear blue sky above. Dinner was held under a large tent, sporting banners, peaks, and sloping sides, that might have been staged in the sands of Arabia. Michelle Vollman, of *La Petite Pierre* in Madeira, supplied an excellent dinner, delivered by 24 servers who arrived in a school bus. A band from Atlanta arrived in their own bus at noon and departed at 3 a.m., leaving behind the most professional presentation of inspired dancing-music most of us ever witnessed. The weather was perfect, the farm looked beautiful, the guests were so appreciative, and spirits were high.

The only slight mishap to a busy weekend was tables for the hog roast, which were supposed to arrive three days before, showed up 15 minutes after the event started. That raised blood pressures of a few people, but most didn't notice, and the weekend flowed perfectly thereafter.

The weekend was nothing short of spectacular, as I have never experienced before. Of course, when one's daughter is married, the emotional impact is high. Our entire greater family, save one nephew, was present; buildings and fences of the farm had been renovated; our home team had worked hard and effectively in preparation; pastures were green from the best growing-season ever; cows were calving next to the tent; the weather was clear and generous; guests were full of gratitude and affection; and dear Susan is my partner in life! What a collection of forces being expressed in a great melodic symphony over two days. We have never held such an occasion as this and will not again.

But I kept wondering, "how does all of this happen, where does it come from, how did it arrive?" We can't live in spectacles, like this, for more than a moment, but we can reflect on how one develops a sense of triumph, of fulfillment in life, worthy of celebration. Triumph cannot be purchased. It only really arrives over the long run: step-by-step, brick-by-brick, good-decision by good-decision. It happens the hard way, by doing one's homework, by expressing one's values, by daring to be who one is over a long period of time. It takes courage to do this, which is often lonely and difficult, but it claims one's virtue and enables one's voice to surface. Finding one's voice is a mysterious process, but it starts with these small steps of courage.

Coupled with steps of courage would be expressions of gratitude. There is so much goodness in the difficult path of life, that if one recognizes and affirms them, the soil of one's life becomes watered and ready to sprout its unique expression.

During the farm tour on Saturday morning, we discussed how grassfed meats support the environment, by pulling carbon from the atmosphere and storing it deep in the soil though roots of grass plants. Most of the

guests were from out of town and don't have farms of their own to do what we do, but they can go to local farmers markets to buy grassfed meats, not only for their nutritional benefit but also for the environment's. We encouraged that.

This led us to feeling gratitude for our own customers, for all of you. We already have wagon loads of customers who are supporting us, the environment, and their own well-being through the purchase of our grassfed meats. We are deeply grateful for you and are proud to have the relationships we do with you. It has only come over time: week-end-by-weekend, bite-by-bite, slider-by-slider. Our out-of-town visitors made us all the more appreciative of you, and we look forward to returning to the market on Sunday.

Friday evening's guest of honor.

I have been meaning to pose the following: would anybody be interested in buying half a beef wrapped in paper, rather than vacuum packaged? We have one customer interested in such a half but we need another to match him. We will take the beef to a different processor who only wraps in paper. Our current processor doesn't use paper anymore.

Yesterday, I picked up four of our hogs, packaged and ready to go. The pork chops look great.

We stand in celebration of Mary, Leighton, and you.

Collecting Seeds
October 21, 2016

Wild seeds are drying in sheets and will be spread this winter.

Seeds for Wool Grass on the left and Elderberry on the right have been painstakingly collected this past month by Naturalist Kathy Kipp. They, along with False Indigo, Nine Bark, and Buttonbush, are drying in thrift shop sheets, by hanging in the wind. They will be cleaned, sifted, and then broadcast in December by Kathy and Jacob Bartley onto bare patches of soil, in 100-foot buffer zones surrounding our wetlands.

Collecting and planting seeds is a hopeful and essential activity. The trees below were planted 8 years ago in our wetlands and are now 20–30 feet tall, growing about 3 feet per year. Think of the good they are performing —sequestering carbon through photosynthesis, creating habitat for wildlife, cleaning water, purifying air, creating employment for naturalists, generating revenues through wetland mitigation, all while pulsating with spirit. They emerged from 2-foot tall seedlings that sprouted from the almighty seed. The power of seeds is amazing and lies at the heart of the natural cycle, this restored wetland, and our own dream for sustainability.

One plants seeds of hope to realize dreams, to grow an ecosystem of visions. Many seeds must be planted for a few to emerge and for even fewer to germinate. The ground must be fertile and moisture plentiful for abundance to take root. But abundance will take root if one is faithful to one's dreams and liberal with seeds of hope dispersed upon them.

First, however, one must collect seeds distinct to the self, requiring intentional acts of selection. Kathy knows the difference between Nine Bark, Wool Grass, and Elderberry, and will be discriminating in how and where each is planted. One must be thoughtful about what one plants, so one's true visions will emerge to grow, perhaps at the rate of three feet per year. If one is careful, persistent, and hopeful, one will realize a boulevard of vision, like the picture above on the right. We refer to this as Bartley Boulevard, in honor of Jacob, who is the architect of our wetlands. Such a boulevard awaits anybody who cares, persists, bears hope, and is willing to collect and plant appropriate seeds.

A meal of pork shank, polenta, cannellini beans & kale, baked apples, and a late-season, tomato salad. The shanks were cooked 14 hours at 200 degrees, and were great.

The growing season is drawing to a close and outdoor farmers markets are winding down. This presents a challenge for those of us with bills to pay year round, so we hope you will stay with us one way or another through the winter, as we have product available all year long.

May we each continue to collect necessary seeds, that will grow into the vision we imagine for ourselves.

Wildfire
December 1, 2016

Wildfires of the southeast paid us a visit two weeks ago.

We have enjoyed beneficial rains every week of the spring and summer, making for an ideal growing season, until mid-August. Since then, we have not received more than two inches of rain. Two Saturdays ago, the wind was high, the grass was dry, the leaves on the forest floor were very dry, and a neighbor at the top of the hill started an habitual trash fire in her backyard around noon. By 1:00 we could see thickening plumes of smoke billowing into the sky. We could also hear sirens of fire engines approaching. The trash fire had escaped onto dry grass and rapidly advanced into our forest. A level of dread arose, from our distance, as the prospect of a full-fledged forest fire seemed imminent. Fortunately, the neighbor had called the fire department right away, and by the time I arrived at the scene with shovel in hand, the flame was largely contained by heroic firemen. But several acres of understory of our woods had burned and could have spread over several hundred acres. It was a very close call and an alarming event that caught everybody's attention. This brief experience with large fire apprised me how small and vulnerable one is when confronting a serious inferno. Interestingly, the primary tool for arresting the spread of the fire was a leaf blower.

We have not had to deal with forest fires much in the past, but with the advent of climate change and droughts, this may be an increasing phenomenon with which we will have to reckon.

Ashes of debris along old fenceline.

When trauma like forest fires threaten, it is always heartening to take stock of the youth on the farm.

Pork chops, which were put on the grill to be seared, then cooked in a 400-degree oven for five minutes. They are accompanied by a cherry sauce, rapini, cornbread, and Canellini beans. The chops were brined in salt water and sugar for several hours beforehand, adding to their succulence.

Thank you for fighting wildfires of all sorts with us!

Buildings
December 15, 2016

Buildings are all womanly.

Their roofs are like the flanks of mares, the arms and the hair of wives.
The future prepares its satisfaction in them.
In their dark heat I labor all summer, making them ready.
A time of death is coming, and they desire to live.
It is only the labor surrounding them that is manly,
the seasonal bringing in from the womanly fields to the womanly enclosures.
The house too yearns for life,
and hot paths come to it out of the garden and the fields,
full of the sun and weary.
The wifeliness of my wife is its welcome,
a vine with yellow flowers shading the door.

                                                                      Wendell Berry

Spiritual adviser to many of us on the land, Wendell Berry ascribes womanhood to buildings. He sees in them the strength, embrace, and raw beauty of women. Many might regard buildings as male, being stout and forbidding, as some are. But most barns display a grace and majesty that is indeed feminine, which Wendell Berry articulates so clearly. His view of womanhood is also far-ranging, from engagement in fields where action is undertaken, to standing as buildings with flowing rooflines where protection is found and the future secured, to offering "wifeliness", through which partnership is consummated. This is a full and dignified regard of womanhood.

And it is fitting, for life on a livestock farm is centered around females. Our farm supports over 400 of them! That is a lot of estrogen to consider, but consider one must, if the abundance they offer is to be realized. We are trying to realize improved egg-abundance through the winter, by offering more shelter in the barn, than is possible with the mobile unit.

We recently received an order for a large amount of bone broth from a gentleman who has been experiencing issues with digestion. This order galvanized production of bone broth and chicken stock, as pictured below. It is a long, slow process of at least 24 hours to make good stock of any sort. The resulting nectar enriches all soups, vegetables, and slow-cooked meats.

We also received word that meat we donated to the *City Gospel Mission* on Dalton Avenue was recently prepared and served with much pride. *City Gospel Mission* carefully supports those dealing with homelessness and addiction to reenter society, through a rigorous and extended program. Offering good nutrition is essential to this process. We are

grateful to contribute to a cause so worthy, and couldn't do so without your support of us, making connections through food far-reaching. Chef Antonio prepares legs of lamb.

Making omelettes—a rewarding food at any time of day and in any season of the year. Few foods are so versatile.

May the womanly buildings in your life continue to prepare the future to your satisfaction.

Slow Fat
February 2, 2017

These steers have put on fat slowly. They are ready to be harvested, at four years of age rather than two, which is customary for a finished grassfed animal. We observe that older animals offer more flavorful fat, so we don't worry if they aren't ready as soon as the industry suggests. Some of our best steaks have been from 8 and 9-year-old cows, who have had plenty of time to mature. Another aspect of finishing a steer in 24 months or less is he needs to consume a lot of high-quality feed to do so. We finished our first batch of steers in 20 months by feeding them baled alfalfa for about six months. It worked great, but the cost was too high. Another option is to grow annuals, like Sudan grass, turnips, wheat, cereal rye, and graze those for finishing. But those are costly to grow as well, requiring considerable equipment for tillage or use of herbicides, both of which we try to avoid. Planting annual crops means forgoing perennial forages. It is perennial forages that protect the soil and offer opportunity to build organic matter at least cost. Over the past two years, organic matter in a number of our pastures has increased 50%, by implementing a tight grazing plan with our growing herd of cattle. As we have discussed, organic matter is the gold standard for providing nutrients to plants and resilience in drought. In the one field we plowed for organic corn, organic matter dropped by 50%.

So, the simplest and perhaps the least costly solution to finishing grass-fed beef is to go slow—let those steers take their time coming to maturity on average grass. Slow fat is working for us. It is like slow food, slow money, and slow haste, all of which seem to stand up to the tests of time.

This beef tenderloin, grilled rare, was fabulous.

For those of you who enjoy the alternate reality of the Super Bowl, we can provide chili or sliders for your fare, that are guaranteed to please the crowd. Let us know in advance if you need a quantity of either.

May slow prosperity come quickly to us all.

Man's Best Friend

February 16, 2017

Working dogs are a marvel.

Bo and I have been sorting and moving livestock regularly this past month. We are discovering teamwork for the first time, really, as he arrived here two years ago at the same time as Brendan, who thereafter worked with him. Bo and I are getting to know each other and it has been a pleasure to do so. Working as a team is an imperfect process, as we all know, and he tests his handler regularly. He likes to advance way ahead of the handler, creating disconnection, and has to be continually called back. But he is smart and eager to learn, and responds readily to both positive and negative reinforcement.

The other day we sorted yearling lambs, destined for the market, from ewes who are unexpectedly having babies. Not wanting to separate newborns from their mothers by sending Bo to round up the group in a stampede, he and I walked together to the flock in the field, and were able to surgically leave behind mothers with newborns, while taking the rest to the sorting pens. Once in the sorting pens, Bo showed an understanding of pushing the flock forward, in a way he did not a year ago. Some dogs are good in the field but not in the sorting pens. Some are good in the sorting pens, like Nick, but don't have the discipline to stay in the rear of a flock in a laneway to drive it forward. Bo does all three, better than any of our previous dogs – gathering in fields, driving down laneways, and pushing in sorting pens.

The Border Collie is one of the smartest canine breeds, as its genetics are closest to the wolf, from whom domesticated dogs descend. A highly skilled handler of border collies demonstrates absolute control, typically through whistles, which can be heard at far distances. Such a master can induce a dog to dance a ballet, and it is beautiful to behold. But back on our farm, we resort to a few basic commands, issued by voice, which suffices in general, though is not always pretty. The commands are: "come bye" (swing to the left), "way to me" (swing to the right), "lie down", "walk in", "stay", and "that'll do." When calling signals to a dog, it almost sounds like one is calling a square dance.

Having a dog that is good with sheep is one thing but having one that works cattle as well is another. Because of the difference in size of beast, most dogs just refuse to deal with cattle. We had two such border collies – Jazz and Dally. Nick was willing to wade into cattle, and so is Bo, which greatly compounds his value.

The other day, we were moving steers from a distant paddock to the barn to sort several to be harvested. Once we had done the sorting, the group of 40 animals were in a holding lot and weren't finding the open gate by which to return to the laneway. This group was feeling frisky and began running around the lot. So, I sent Bo to collect them, hoping for the best but not thinking that would really be successful, because a herd of charging bovines is hard to influence. But intrepid Bo swung around to the front of the lead steer who was on the move. He opposed the steer, then gave some ground, slowing the group. The steer charged Bo; Bo gave more ground, slowing the group further; the lead steer then charged a third time, and, in response, Bo leaped in the air toward the steer with canines snapping right in front of the steer's nose! That lead steer was shocked. He and the herd following came to an abrupt stop, reversed direction, and willingly moved toward the open gate. It was amazing to witness a 70-pound dog reverse the momentum of 40,000 pounds of bovines. Nick wasn't strong enough a dog; he would have retreated after the first charge. Not many dogs are that strong. It requires unusual courage to confront thundering hooves, such as Bo demonstrated that afternoon.

But he is not our only working canine. We also have Maremma guard dogs (from the province of Maremma in Italy), who keep coyotes at bay, in protection of sheep and laying hens. Those dogs are: Coquie, Kentucky,

Max, and Abie. Because of them we are able to enjoy the cacophony of coyote howls at night, without worry.

Ulysses

Coquie

Kentucky and Abie

Bo

We periodically receive tributes for our food. The following is recently from Catherine, regarding Chicken Stock:

We had the most delicious chicken noodle soup made with Susan's broth plus a grocery store rotisserie chicken. It wasn't 100% perfectly homemade with that chicken, BUT it was almost and more importantly, it came together in less than 30 minutes. Susan's broth gave a rich multi-layered flavor that barely needed any extra herbs or seasonings. The kids ate it and we had leftovers for the next days' lunch.

We really appreciate this kind of feedback. It makes the journey worth the great struggle. To that end, we have recently set up a Yelp account, where such reviews can be collected to create a deeper picture of the work underway. Go to Yelp, search for Grassroots Farm & Foods, and post your review. This will help us build presence in the digital marketplace. We would be grateful for your efforts in this regard.

Several weeks ago, we witnessed a spotlight of sun upon the grey horizon, inviting us forward to higher ends, not unlike the spotlight provided by a review in Yelp.

Susan chooses her mentors more selectively than most of us, with Mr. Lincoln being at the top of the list. We thus have a dog, Abie, and we thus celebrate Lincoln's birthday, as if it were Christmas. The repast acknowledged him on his day.

Note the stunning lettuce-rose, from Farm Beach Bethel. The grassfed lamb loin chops were as delicate as wild tuna. Backyard Orchard provided local fresh fruit, with which to make delicious baked apples. Polenta, sweet potatoes from Elmwood, and creamed kale, also from the Mancinos, made for the vegetables. Abe would have been proud, though it appears he was not all that interested in food. Nevertheless, the celebration was in tribute to a great man.

Blessed be *man's best friend*, without whom life would be less.

Flood Waters
March 2, 2017

The power of water visited us Tuesday night.

Four inches of rain fell upon saturated soils, culverts bulged, creeks rose, and fields flooded. We were reminded once again of nature's ready ability to transform the environment. It is humbling to wake up to such a transformation, especially when one has 500 animals to provide for. One feels small and powerless.

But we have learned over the years how to cope. Landis' dairy cows were under roof, as were his first newborn calves. His bulls were marooned on an upland island, so they were fine. Our cows were on high ground, as well, as were hogs and sheep. Laying hens are under roof fortunately, as they would have suffered from such exposure. The only challenge really was taking grain to the hogs, who were across the creek on a high knob among trees. Had to use the tractor to ford through two feet of water, but it worked.

By noon, the flow through the culvert had reduced by half, and flooding over the road by Landis' house had receded. It comes quickly and goes quickly, but when it comes, it comes with a fury.

This was a reminder, yet again, that water is a determining factor of life. The more we work with our farm, the more we realize its success is primarily contingent upon managing flow of water. That flow involves: wells, springs, county water, underground waterlines, underground tile lines, ditches, creeks, and organic matter. Wells, springs, and county deliver water to people and livestock; tile lines, ditches, and creeks transport water away; and organic matter retains it in soil. We are additionally exploring installing control-valves in tile-lines to sub-irrigate pastures. All of these components work together to keep water circulating past, through, and within the farm, in sustainable fashion. This is an intentional process, full of design discovered over decades.

Water is a fascinating topic. Nothing catches our attention more than having too little or too much of it. We were at full alert Wednesday morning.

While driving back from taking grain to hogs through floodwaters, I noticed atop a bale of hay, a large, black turkey vulture perched in ominous fashion. One is always concerned when witnessing a turkey vulture in the vicinity, wondering where lies its carrion. Newborn lambs in the adjoining field are also vulnerable to such a predator. While conceding he was rather magnificent, perched 15 feet high upon the bale, I witnessed something I hadn't seen before. One of the guard dogs, Kentucky, in with ewes and lambs, noticed him from afar as well, and began barking, and then broke into a run at full tilt toward the vulture. The two other guard dogs took notice and joined the chorus and pack. So did the turkey vulture take notice, and soon enough alighted for more favored circumstances than tangling with three barking 100-pound canines in full assault.

What was interesting to experience, for the first time, was that our guard dogs survey and act upon aerial threat as well as terrestrial.

We recently enjoyed a delectable dinner of ribeyes, Mancino carrots, rapini, mushrooms & peppers, and beans.

In the power of water, may our thirst be quenched, our roofs tight, our fields green, and our basements dry.

What's in a Name?

March 23, 2017

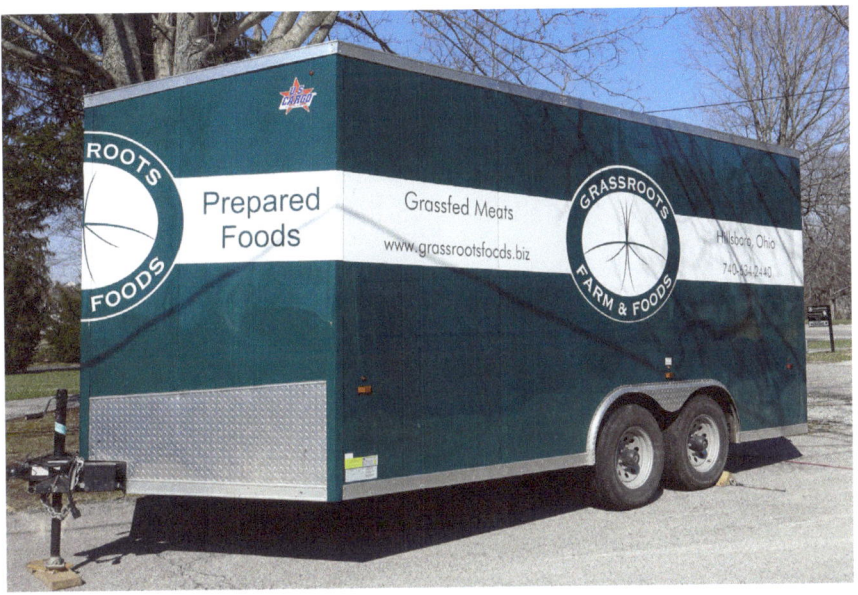

Our name was just inscribed upon this trailer.

Why does one do this? Isn't it awfully brash to promote oneself like this? What would one's grandmother say about being so demonstrative in public? Is it really necessary? And how important is a name, anyway? These questions arise, as the signature event of the week was installing this large banner around our large trailer.

One of the most effective claims to name and advertising I recall from decades ago was promoted by *Buddy's Carpet Barn*. Buddy would take 30 seconds to explain, in growing breathless enthusiasm, the excitement around his low-priced carpets. All he was selling was price, a mere commodity, but his message was so enthusiastic it was infectious. He was successfully brash.

Our personal names are usually given to us by parents. Sometimes they fit the recipient, creating welcomed lifelong affinity. Other times they don't, forcing definition upon a newborn that creates discomfort ever after. Some nicknames are so descriptive and fitting, they won't shed themselves, no matter how hard the afflicted try to cast them off, like: Jocko, Tugger, Buffy, or Yesod. Many women have been willing to change their last names, when entering marriage, accepting the mutability of identity. In the nineteenth century, notable women authors, buried under the weight of sexism, assumed pen-names, like *George Eliot*, by Mary Ann Evans. Latin cultures ascribe four or five names to an individual, to recount full lineage. Some individuals want to free themselves from cultural confines, and stridently claim only one, as in *Madonna*. In Indian culture, first and last names are often repetitive. In Anglo lore, if one were born into families of: *Montague, Capulet, Hatfield,* or *McCoy,* the only way to escape a bloody future was through the defiant powers of romantic love. Members of aboriginal cultures assume new names with each stage of life, reflecting the complexity and richness of great journey. At the same time, names usually identify one's historic tribe, which can be emotionally centering.

At our farm, we engage in this awkward discourse with names as well. All of our fields are numbered, which are names in themselves. My aging father remembered the numbers of every field, and loved to recount the history of each, as we drove by. Susan doesn't resonate with tedious sequential numbers, much preferring descriptive terms. So, Fld 5 is also Sardinia, Fld 27 Snake Field, Fld 29 Turkey Vulture Field, and Fld 23 Caribbean. We have stopped naming most livestock, learning the hard way that doesn't work, but the eight dogs have their own affectionate designations, of course. The legal entity that owns our land has a different name from that producing and marketing grassfed meats, the legal name of which is also slightly different from that under which we do business. The creek flowing through our farm, saddled with an unimaginative Anglo name, succeeded the designation by Native Americans over thousands of years. So, the topic becomes rather complex, reflecting the challenge of ascribing static names to dynamic forces.

In the end, though, one has to claim who one is, if one is going to participate and compete in the world. The best way is through one's actions. The name one assumes, however, is also important, as it represents those actions. We have chosen *Grassroots Farm & Foods*, to reflect how the roots of grasses stand at the heart of all we do and stand at the heart of good health, in many ways. So, with deference to our genteel grandmothers and with a lesson from Buddy, we have summoned the courage to impose our name upon the side of our trailer.

May the world behold for an instant who we are.

Orecchiette pasta, Grassroots Italian sausage, and rappini from a Beach on a Farm in Bethel conspired to create this hearty meal.

May the names we choose honor the past while embracing the future.

Walk in the Woods
April 20, 2017

The onset of Redbuds is a harbinger of riches to be revealed in the woods in the year ahead.

Nature is obviously coming awake, as grasses are growing, trees are turning green, birds are migrating, and wildflowers are emerging. Susan and I took a walk in the woods last weekend to look for wildflowers. We are typically too busy for leisurely strolls, but the weekend off provided welcome opportunity.

We were primarily looking for Trillium, but also found Trout Lily, Spring Beauty, Hairy Buttercup, and Phlox. The woods have not yet become dense with green, so it is an inviting moment to witness contrasts between winters' brown cast and springs' emerging colors. Morels are imminent, but would not reveal themselves. Orange fungus on decaying limbs did, however. Best of all, a majestic, decaying, oaken log gave succor to lungs and legs, from which to behold the tapestry.

The woods at this moment are like a museum showing a few of its prize artifacts, before the show commences and crowds descend. It is a privilege to witness nature's first pageantry.

One situation in which bleached white flour stands proud is within a stack of pancakes! These were heavily accented with Ontario maple syrup, Irish butter, and blueberries. Grassroots woodlot bacon accompanied to legitimize this sumptuous indulgence.

We are restocked with bacon, should you seek some for breakfast feasts, sandwiches for lunch, or omelettes for dinner.

We just received this testimonial from Dr. Gary Huber, who recently bought a side of beef from us.

As promised some report on your product. We grilled the strips and the flank for Easter Sunday and it was AWESOME!! Great taste and surprisingly tender for grass fed meat. A thing of beauty.

I can give a testimonial for your farm any time you need it. Great work Drausin & Susan. Greatly appreciated. I hand your card out to all who are interested in getting a side piece of beef.

With gratitude for the ongoing walk in the woods with customers and for the brief one last weekend with wildflowers.

Marketplace
May 4, 2017

Marketplaces stem from ancient tradition.

With the advent of agriculture in the Middle East around 10,000 BC, surplus commodities enabled producers to begin bartering for goods to improve quality of life. Marketplaces began to be formed to facilitate such transaction, and amazingly they continue to this day, throughout the world, in nearly identical form. Vendors arrive at dawn with surplus goods in hand, set up shop with tent and table, and spend the day exchanging currencies, hopefully for the better by dusk.

The *Agora* of Athens, was a gathering place, built in the 6th century BC, not only for the exchange of goods, but also for military mustering, political discourse, and philosophical debate, serving as one of the seedbeds of democracy. *Samarkand* (above) on the "Silk Road" in Uzbekistan was built in 700 BC by the Sogdonian peoples. The Sogdonians allowed any foreigner to settle in the vicinity, as long as they abided by laws-of-trade. Greeks, Turks, Mongols, and Chinese reveled in the prosperity that developed at that site, and each made Samarkand capital of their empires over time. *Trajan's Market* in Rome (first century AD) was notable for being five stories tall. The *Grand Bazaar* in Istanbul was built in 1455 AD to include 4,000 shops and 26,000 merchants. *Timbuktu* (Mali) became the largest marketplace in Africa, as it stood at the intersection of Arab nations to the north and African empires to the south, "where camel met canoe." In the 1600s, the *Bazaar of Isfahan* in Persia was constructed to include one mile of continuous vaulted ceilings. This Bazaar was so prosperous that Isfahan became the capital of two Persian dynasties over the succeeding 700 years.

These marketplaces stood at the heart of culture for regions they served. Not only were local goods exchanged, but also foreign ones, leading to exposure to diverse elements and broad perspectives. They also served as important locales for social interaction, when travel was difficult and socializing constrained.

And then we have North American marketplaces, characterized by "the Mall." Edmonton, Alberta offers the largest market under-roof in the world, with 800 shops, 20,000 parking spaces, 3 amusement parks, and 200,000 shoppers per day! Imitations of Edmonton have been built across North America, in distortion of a true marketplace. Most of these malls are not standing the test of time, for they lack soul, heritage, and distinction. They are venues of tired mass-production, rather than purveyors of vigorous, local excellence.

Markets that have survived the ages, such as Findlay Market in Cincinnati, plus the advent of farmers markets across the country, speak to an unfulfilled need in American society. That need is for food consumers can trust, produced with excellence and craftmanship. These are old-world values, which have stood the test of centuries, and are finally taking root again on this continent.

There are problems with farmers markets. They are unreliable, due to weather and competing local events, making for a difficult business model. One has to transport excess inventory not to disappoint potential customers, which is inefficient. Too many customers disappear during winter. It is difficult to scale-up the business through this venue of face-to-face transaction.

But farmers markets offer many virtues. It is a great place to start a business and introduce products to people. Customers become friends, as one discusses the implications of nutrient-dense food on life. And one becomes a part of a community of outstanding fellow-vendors and farmers, where invaluable support is extended and received. Susan and I benefit a great deal from these virtues and are grateful for the opportunity to participate in this ancient tradition of the physical marketplace. We accordingly thank Mary Ida Compton and Judy Williams for their effective leadership of the Hyde Park Market over the past decade. Such leadership is a profound contribution to society. We thank our fellow farmers, whose high character and extreme work ethic delivers cherished excellence, for which our culture is hungry. And we thank invaluable customers, whose discerning eye and faithful presence keeps the wheels of old-fashioned commerce turning.

As fast-paced technology continues to dissolve distances and relationships in current society, it is likely that face-to-face interaction over local food will become all the more sought-after and necessary. It appears the ancient marketplace may stand forever, and praise be to it!

Laying hens make quite an impact on pasture. Our new pullets were in the top section for a week. We moved them yesterday to fresh clover and grass. We have spent a few evenings tenaciously transferring somnolent pullets from beneath the mobile coop to the inside, so they learn where home lies.

This meal came from the Hyde Park marketplace, featuring Grassroots pork tenderloin, Walnut Ridge asparagus, and chard and locust-blooms from Farm Beach Bethel. The fruit compote consisted of mangos, blackberry, and honey. It was a fabulous meal. The night before we enjoyed fresh asparagus soup, the base of which was pastured chicken stock. You, too, can dine primarily from your local marketplace, in concert with tradition thousands of years old.

In gratitude that you walk with us to maintain ancient tradition.

Turkey Tail

May 25, 2017

Behold this beautiful tail of a tom turkey.

Most of the month of May is turkey-hunting season in southern Ohio, and this shimmering, elegant tail was recently offered to me by neighbor Kathy. Several weeks ago, she enticed a tom to present itself closely enough in the woods to feed her family on Memorial Day. The colors of the tail are breathtaking in their muted, silken flow of browns, reds, and yellows, which most of us don't witness up-close. Some of nature's greatest artistry seems to be expressed in the plumage of birds.

I asked Kathy to share a few words about the tom turkey.

In Spring, a tom turkey's fancy turns to romance. The brilliantly colored tom gobbles, struts, drums and spits, doing his best to attract dull, unremarkably colored hens. It's the hen that makes the choice if or when to accept his suit. Older more dominate toms try to outdo challengers, often sparring and chasing lesser Casanovas away. After weeks of strutting and flirting with the girls, a tom's weight drops, his wing feathers break at the tips and his tail becomes ragged. A mature tom can weigh 18 to 25 pounds, depending on breeding activity and diet. A hunter lures a tom by attempting to imitate a receptive hen, using a variety of calls to convince him there is one more girlfriend who can't wait to meet him. And then, there is always luck. Only toms or bearded hens are legal to harvest in Spring, for hens are nesting, supplying offspring for the future.

Thank you, Kathy, for your interesting words and magnificent gift. What a beautiful tribute to the abundance of nature and this season of hope.

There is contrast between pasture grazed yesterday and new grass for today. Grass trampled onto the ground feeds microbes which activates the mineral-cycle in soil. Grass that is not trampled, but grazed, sloughs root-matter into the soil, releasing carbon and building organic matter. Each gram of carbon stored in soil attracts eight grams of water. The more carbon in soil, the more water is retained and resilience generated against drought.

Grazing is a powerful tool not only for producing nutrient-dense meat, but also for rebuilding topsoil, upon which civilization depends. Few tools are so useful. Nature has been perfecting this one for millions of years. It is thus prudent to employ it, which we do with your help.

Two weekends ago we enjoyed a great Farm Tour, with a group of 20. We banged through pot holes in trucks, watched Bo shepherd reluctant yearling sheep, observed ewes with newborn lambs, examined two different shade-mobiles, said hello to steers ready for harvest, observed the cow herd with calves, inspected the egg mobile and its guardian, Coquie, saw woodlots where hogs were raised, took a look at wetlands, and enjoyed a sumptuous meal prepared by Susan of: Moroccan sliders, egg salad with Moroccan mayonnaise, baked beans, and homemade chocolate chip cookies. The weather was beautiful, and the day unfolded perfectly, providing opportunity for customers to understand more closely the landscape and processes behind the food we bring to you.

We are transitioning from old hens to new hens, which is challenging our supply of eggs. But the pullets who arrived several weeks ago are starting to lay, so supply is rebuilding. In the meantime, pullet eggs are smaller than full size, but are rich and delicious. A recent lunch of pullet eggs and pepper. Couldn't have been better! We will bring them to the market, at a discounted price, until they reach full size.

May the beauty in a turkey's tail give flight to your dreams.

Summer Solstice
June 22, 2017

The timeless serpent of this mound honors the summer solstice.

The alignment of the head of Serpent Mound, in Adams County, points to the setting sun on June 21. Beautifully undulating coils of the serpent's body align with the winter solstice, the spring equinox, and the fall equinox. This fascinating earthen structure was built around 300 BC, by sophisticated Native Americans, upon land thrust upwards by the strike of a meteorite, rendering it dense with astronomic power. Serpent Mound is over 1,300 feet long and three feet tall, and is the largest serpent-effigy in the world. It is a powerful site, full of concept, dignity, and magic, on par with the cathedrals of Europe and the pyramids of Egypt.

It is managed by *Arc of Appalachia*, is a National Historic Landmark, and is being considered for designation as a UNESCO world-heritage site. This magnificent serpent resides only 15 miles from our farm, the picture of which was taken last weekend.

The moment of the Summer Solstice is potent with affection and promise. At this time, birthday parties glide into endless evenings, fireflies fascinate and decorate, and wildflowers adorn a sense of the future, whether they be Tiger Lilies, Milkweed, Wild Rose, Elderberries, or Trumpet Vines.

Human Treasury is precious, and two significant contributors to us, at this time of year, are Susan's children, Sebastien and Alexandra Hue. They are tireless contributors to the Slider Shack, now with its own tent. They keep the delivery of grilled Moroccan, Vietnamese, and American sliders moving forward expeditiously. We are most grateful for their faithful and heartfelt support. It takes a village, of which they are a part, to do this kind of work.

Sebastien inherited his mother's passion for relentless study of culinary culture, and recently managed to parlay this knowledge into a consulting assignment with Kroger. His nascent marketing company, *BS LLC*, was hired to accompany a buying-trip to Italy. He wrote an extended report, under the pseudonym of Blake Simpson, for those interested in how the fast-pace business of Kroger might intersect with Italy's ancient world of slow-food. His account reflects high intelligence and deep understanding of food and culture.

May the promise within summer's solstice inspire us all.

Tall Grass
August 3, 2017

The wet spring and summer have created an explosion of pasture.

This field hasn't been grazed since April, so chicory and Queen Anne's Lace are in full expression, four to five feet tall. These plants are pretty mature by now, so cows don't pay much attention to them, but they are interested in the fescue, orchard grass, and clover closer to the ground. In pursuing their feed, they trample mature plants, creating a mat of mulch. This keeps soil moist and cool, and brings plant-matter in contact with soil to feed microbes within, perhaps the most important species on the farm. An active population of microbes is at the heart of creating organic matter.

We are in a quandary about what to do with all of this grass. We typically don't mow a pasture until after the cows or sheep have grazed it, so they have a full range of nutrients from which to choose. This includes plants that are mature and have gone to seed, typically up high, providing energy, and plants that are less mature, typically down low, providing protein. After the stock pass through, we usually clean up the pasture with mowing.

In recent years, we have not made hay, because our cow numbers are expanding and we want to save grass for them. But this year, we could have made hay, and still had plenty of pasture available. In April, when one has to decide about hay-making and hay-purchasing, one doesn't know how the season will unfold, so it seemed prudent to take the cautious route and buy the hay. Purchased hay also brings nutrients to the farm, which is a benefit.

Our "dry" cows, whose calves were weaned at 10 months of age, enjoy grazing. The cows will be calving in mid-September, and we are deciding in which pasture they should be located for this important time. The 30-acre-pasture of choice hasn't been grazed since early May, so it will be about 4 feet tall at that time, if we leave it untouched. Will the feed be good enough for lactating cows? Will we be able to find the newborns in the tall cover to eartag them? How will we walk through growth that tall and entangled to set daily breaks of grass? We aren't sure, but the thought is to mow half the field now, so it has time to regrow sufficiently before calving, and then mow strips in the other half for daily breaks. We will observe the difference in outcome, if any.

One of the concerns about grazing mature feed is whether the animal is receiving enough nutrition. The way we monitor nutrition is by watching the triangular "rumen-fill" on the left side of the cow in front of the hip bone. The other indicator is to watch the manure. If it becomes too dry and hard, cows are receiving too much "dry matter" and not enough protein. Both of these indicators tell us the cows are currently doing well in the four-foot-tall weeds and grasses.

The picture on the left shows a good manure paddy—an all-important currency on a grass farm. The picture on the right is pasture where weaned calves are now grazing, which is the best feed on the farm.

We have started another batch of Berkshire hogs, which should be ready in about 4 months. In the meantime, we are out of bacon, which is a serious omission for our bacon-lovers. The challenge is that bacon constitutes only 10% of the animal, which makes it hard to gear up for sales of bacon, until outlets are found for the other 90% of the animal. To that end, Susan is creating a recipe for Pork Ragout, which includes ground pork, soy sauce, ginger, pepper paste, and other sundry herbs and spices, creating a delicious and easy ragout to go atop rice or noodles. Hopefully, this fall, we will be organized enough to introduce this tasty new product. If we find some success with it, we will buy a sow or two and begin raising our own piglets.

The bouillabaisse Susan prepared for her birthday tasted otherworldly—an exotic combination of the ocean and heavens. It was truly extraordinary, by almost any standard. This fish stew stems from Marseilles, France, but originated further along the Mediterranean in Greece. What is exceptional is the concentrated fish broth, that delivers an upward spiraling ecstasy, which land-based foods cannot provide. Alexander Rostov would have been proud.

Lest you are concerned Susan's cad-of-a-husband just showed up on her birthday to eat godlike food, you can be reassured her negotiating skills, refined daily in the courtroom, exacted a handsome price for that privilege. She shamelessly looks forward to his "making good." Oh, the trade-offs of life.

Fresh peach pie, that tasted as good as it looks.
May your grass grow tall and your rumen be filled.

The Long Road
August 16, 2017

The road to dignity is long. And so are the roads on our farm which need to be maintained! We probably have 4 miles of laneways to keep operable. The most important variable to effective roadways is drainage. Our farm is low lying in many places, so drainage doesn't come naturally. In spots such as this newly poured gravel, there is no drainage ditch to access. Instead we put down "geotextile cloth" to support the gravel, which keeps it from continually disappearing into underlying mud.

Laneways in good repair enable us to navigate around the farm readily, and complete our business. When in poor repair, movement slows down, damage is incurred to vehicles, and the cost of doing business rises. It makes one appreciate the challenges developing countries face with poor road systems. It is hard to market goods when producers are not provided ready passage to marketplaces.

In like manner, pathways provided by the internet have recently become indispensable, perhaps more so even than the physical highway system. This is hard to fathom, but the access created by the internet is unmatched in history. People in remote locations now have access to communication, education, and commerce, in magnitudes previously unimagined.

Numerous remote locations, however, are still without full service from both roads or internet, creating competitive disadvantage. Our farm is one such location, as are many rural neighborhoods throughout Appalachia. Hopefully, someday soon, our politicians will find the courage to pass an infrastructure bill, creating even more opportunity for those who seek it.

Most worthwhile roads in life are long. The journey to success and dignity requires drainage, support, surface, vision, and time. This includes the great challenges of: marriage, child rearing, friendship, starting businesses, and developing careers. The journey never seems to come to an end, as we are always creating new dimensions of ourselves, with every bend along the way.

It is interesting to note most straight roads undulate, subtly and rather beautifully, conceding that "straight ahead" does not exist in nature. Nor does it exist in our personal journey, despite how hard we try.

Similarly, our country is traveling down a long and remarkable roadway. It seems, however, we are recently running into over-sized potholes. But those can be fixed, with proper drainage, support, resurfacing, and faith.

One large pothole that breaks a lot of axles, and thereby invites drainage, is the concept that a single race can be supreme. Supremacy does not exist in nature. The only thing that reigns supreme in nature is connection, among all species. When connection is broken, instability quickly follows. The mighty grizzly bear is weak without small berries. Sustainable biological systems depend on diversity of species above all. And sustainable human systems do as well. Nature provides a time-tested model for behavior, which we humans would be wise to emulate. Our dignity depends on it. Fortunately, the road is long, and dignity awaits those who persevere.

These sirloin of beef kabobs are great for quick grilling or stir-fry.

The cow herd awaits the passing of a long, hot day.

With gratitude that the long road can bring dignity for us all.

Soles of Our Feet
August 24, 2017

Wendell Berry reminds us the best fertilizer for land is footsteps.

It takes a lot of footsteps to install these nets around our sheep, and move them every three days. Doing so keeps the guard-dogs in and coyotes out. It prevents sheep from "backgrazing" and infecting themselves with parasites. It keeps the flock on a constantly high plane of nutrition. And it produces the cleanest lamb imaginable—the Midwest's version of wild-caught fish.

We do not treat our sheep with anything more than this management. They are as close to wild as a domestic animal can be. For producers of lamb, the issue of parasites is the hardest to address. Ninety-five percent of lamb-producers, even those raising grassfed meats, treat their flock with dewormers. We followed this standard protocol at the outset, as advised, until we became wise to Wendell Berry's counsel. If we are willing to walk enough, we can solve the problem of parasites, without resorting to pharmaceutical medications. So, now we move the flock every three days, and don't return to the same spot for a minimum of 90 days. This works, but the solution requires a lot of time and labor, far more than any other of our enterprises and far more than most reasonable farms are willing to invest.

Over time, we are discovering that Wendell is right. The more we walk upon the land, the better we know it, and the more it responds favorably. Moving these nets is more awkward than hard, requiring patience. Once one learns how to keep the feet of the posts from becoming entangled with the nets and the nets from becoming entangled with one's own feet, the process of erecting and taking down nets becomes a meditation of sorts. One is continuously walking back to pick up nets from the last move and moving them forward to the next.

Back and forth, back and forth, one net at a time, with feet always upon the ground. After a while, the meditation takes hold, and the net-minder finds himself engaged in wordless conversation with the land, through the soles of his feet. Reverberations in the soil are of such a low frequency they can't be heard with our ears, but they can be through the soles of our feet. We can feel how soft the soil is, or hard, or wet. We can tell if it is in distress or is moving into abundance. Soft-soled boots enable the sound and feel to resonate all the more and barefoot is best for listening to the land.

Children and teenagers know the sensory satisfaction derived from walking barefoot. It is not just the pleasure of feeling grass that is rewarding, but also the unfamiliar sensation of receiving vibrations from the ground through the bottom of the foot that stirs the soul. These are grounding experiences, that put us in touch with profound forces.

When moving nets and engaging in low frequency discourse with vital soil, the ardor of the task is met and can be surpassed by the inspiration received. As one's steps are fertilizing the ground, soles of the feet become active conduits between one's inner self and depths of the earth, ever provoking one's spirit to higher planes.

On Tuesday, we enjoyed a visit from 30 members of the Ohio Ecological Food and Farm Association (OEFFA) for a farm tour of four hours or so. We showed them the entire operation: market trailers, shade-mobile and weanlings, sheep, wetlands, organic dairy, commercial kitchen, herb garden, cows, and laying hens. We did not see feeder pigs, because they were out of the path.

The pigs have graduated from the holding pen for training to electricity, and are preparing to go to their first woodlot.

Our Mennonite neighbors, who raise broilers for us, have started a batch of White Broad Breasted Turkeys, for Thanksgiving. The turkeys are being raised just like the broilers, offering exceptional taste, and will be frozen at 15–20 pounds. If you would like to reserve one, please let us know.

We are smoking four capons today to be picked up this Sunday at Hyde Park. These are great half or whole, but also make for superior chicken salad, mixed with Susan's homemade mayonnaise. Let us know if you would like a smoked capon. We are going to include a boneless leg-of-lamb in the process, so that will be available for anyone who enjoyed those we smoked at Easter.

We had a great time at Blackberry Farm, near Nashville, listening to Emmylou Harris and Roseanne Cash. They are dignified women with a lot of talent. It was quite an experience to witness them up close.

The food at Blackberry is all made from scratch, and is excellent.

But there is no place like home, and last night Susan's Soulful Kitchen produced: roasted chicken legs n' thighs, roasted Italian plum tomatoes, fresh broccoli, fruit salad, and, best of all, mashed potatoes!

May the soles of our feet always serve as active conduits.

Lyrical Lyra
September 21, 2017

Our cows and pastures are dependent on the surrounding forest.

The forest of the hillsides provides ecological stability and serves as a protective mantel for the valley below. Its roots prevent erosion of soil and store water during times of plenty to be released during times of scarcity. Mature trees release several hundred gallons of water a day into the ecosystem, for the benefit of its community, including nearby pastures.

One of the fascinating aspects of forests is the fungal growth they foster in soil. Roots of trees typically extend twice the distance of the crowns, intermingling extensively with each other. Symbiotic to this root system are fungi, which attach themselves to the roots.

Fungi develop microscopic filaments, known as hyphae. One teaspoon of forest soil contains miles of hyphae, densely penetrating the ground in every direction. These hyphae operate like fiber-optic cables, transmitting signals to trees throughout the forest, regarding threats such as: insects, disease, and weather. Ecologists refer to this newly discovered phenomenon as the "wood-wide-web". A single below-ground fungus of an undisturbed forest can cover many square miles, generating an extensive communication network.

Fungi are beneficial to pastures as well, and one can't help assuming those of a forest provide support for and connection with those of a pasture.

Forests seem like nature's most beautiful garden. In southern Ohio, nearly all fields were once wooded, as is the case for our farm. And those fields are constantly wanting to go back to woods, to return home. If we didn't mow them periodically, they would be dense with woody species within five to ten years. So, we are borrowing our pastures from the forest, in a sense. For this brief moment, we are its stewards.

Forests are a model of true community. Mammals, birds, insects, amphibians, and fungus thrive together in their understory. This biodiversity and interaction leads to unparalleled stability and richness, creating a sustainable society.

Trees that are aging shelter the young, and the young grow strong to assume leadership. The sapling of a beech tree will grow in the shadow of its parent for 70 years, and only be ten feet tall. But when the matriarch fades and falls, the sapling is ready to spring into sunlight and growth. One can clearly witness careful nurturing going on in the forest. Over the centuries a forest takes to tell its tale, nuts fall, seedlings sprout, saplings take root, and trees mature.

One of the fascinations of the woods is the sounds it offers, the song it sings, emanating from its branches, leaves, birds, and wind. One can listen endlessly to its uneven melody, becoming transfixed.

As of three hours ago, a new seedling has sprouted in our very forest. Her name is Lyra, and she comes out of the forest's magical song, as our first grandchild, and is thus Lyrical Lyra. Her dignified mother and honorable father will nurture her, so she may join in the chorus and sing of the wild, from whence we all come. The wood-wide-web is busy sharing the news. We cherish her already, as she will lead us into the future, upon our slopes.

Last week a cow gave birth to twins – a bull calf and heifer calf. We were able to habituate the mother to nursing the heifer calf but observed the cow probably did not have enough milk to raise twins. Sometimes demand creates supply, but that did

not seem to be the case for her. So, we are bottle-feeding the heifer-calf, for the time being, and have found a home for her on a commercial dairy, where they have plenty of excess milk from treated cows. The picture is of the return trip to the cow herd, with bull-calf in sled and mother following.

This recent meal featured brined pork chops, fresh lima beans, and the last of corn on the cob. It was exceedingly good!

If you find yourself wanting to grill this fall, we recommend our Shortrib Burgers as they are always a success.

As night falls, and this communication is about to go forth, two packs of coyotes are calling back and forth in our valley, in resonant chorus, howling with intensity from unknown canyons, proclaiming the arrival of Lyra, and our hearts swell in gratitude.

Food & Politics
October 19, 2017

Our farm tour last weekend revealed a collection of cultures convening over food.

Irish-Americans, German-Americans, Mexican-Americans, and India-Indians gathered together in a remote corner of Appalachia to celebrate the growing and savoring of nutrient-dense food, on a beautiful landscape. It was a remarkable cultural event, unlike any we have experienced on this farm. Many widely different perspectives quickly found unity over something essential – healthy food. Such convening seemed effortless and was enriching and gratifying.

One couldn't help wondering—maybe something unusual is going on, something worth taking note of, in this gathering. How could this redeeming experience be magnified and expanded? Perhaps we should consider creating a political party centered on food. Wouldn't that be an interesting idea and a unifying focus? Wouldn't it be great to have a political party concerned about uniting rather than dividing, and what is more unifying than food?! That was the brainstorm of the day, as we stood around enjoying each other. So, beware reds, blues, and greens, the party of food may soon be emerging to unify this nation! And unity starts with the soil. Thomas Jefferson would likely approve.

Without a doubt, the star of the above event was its youngest participant—Shiv, age 4 weeks. Shiv's great grandmother owns and manages a farm in India, at age 99. His grandmothers are both exquisite cooks, presenting ancient food that bursts with flavor, complexity, and balance. His agrarian and culinary roots thus run deep. However, Hollywood already has its eye on him, for obvious reasons. But it is my prediction he will combine good looks with a commitment to food, and will rise from the carriage to lead the *Soil Party* someday. Keep your eye on this young man.

An entertaining outing of the farm tour was a visit to the woodlot, where 10 Berkshire feeder pigs are thriving. We also visited cows, sheep, laying hens, and watched Bo herd a group of 30 heifers.

While cows have crossed the drainage ditch to greener pastures on the previous page, a few calves are resisting the crossing. Mothers have been coming back to nurse, not forcing the issue. This afternoon, we managed to induce the last eight or so calves to get their feet wet and cross the ribbon of blue. Doing so took some struggle.

We have discussed grazing tall. One of its virtues is allowing the milkweed plant to express itself fully, upon which monarch butterflies depend.

Steak salad, with slices coming from a ribeye, but would also work great from a strip steak or sirloin.

Implicitly yours.

Pursuing Excellence
October 26, 2017

Pursuing excellence is almost a punishing venture.

Excellence is always difficult to achieve, but perhaps nowhere more so than in human relations: raising children, nurturing long-term partnerships and friendships, caring for elders, working through conflict, and having difficult conversations. No simple formula presents the pathway forward, which is discovered one increment at a time, like navigating in fog. The process can be unnerving, for reefs are usually hidden. But having a strategy by which to navigate, and trying to stay consistent to it, despite fog, does yield results over time, sometimes long periods of time. Those results, patiently awaited, generate a glow of satisfaction, the reward of work well-done, a sense of approaching some sort of excellence.

We seem to seek punishment at our farm, as we always strive to do better. The most demanding arena is proving to be the kitchen. Preparing food so it delivers the same result every time is a very challenging process. Ingredients need to be consistent. The recipe has to be clear, so it can be followed exactly. "A pinch of this and a pinch of that" doesn't translate well. Protocols for cleaning and preserving are essential to final results. It is an unforgiving process, designed to earn the trust of customers. If conceived and executed properly, excellence is approached. Our Bolognese Sauce is an example. This excellence is a testimony to intricate teamwork between Susan, the creator, and Beth, the implementer.

As our calving-window concludes within a span of 45 days, we feel good about the process and results. All calves, but one, have been ear tagged, identifying the mother, year of birth, and sex of calf. The cows are receiving two breaks of highly nutritious grass per day, along with plentiful fresh water and dried seaweed or kelp. Precious pasture is not being wasted, calves are romping, and cows are in good body condition, in preparation for breeding in December and a long winter ahead. We feel a glow of satisfaction at these outcomes thus far.

An important part of excellence is making mistakes. If we aren't making mistakes, we aren't testing ourselves and moving forward. Excellence comes, in part, from figuring out how to benefit from setbacks, without undue cost. We have had two pieces of equipment stuck in ditches and mud, of recent, in the course of important work. We were able to extract them and complete the task, at minimal cost, with lessons learned to make us better.

Pigs in the woods, sheep on the hillside, dogs on the porch, and dairy cows in pasture each delivers a glow of satisfaction, implying degrees of excellence.

Perhaps the ultimate excellence is being humble to one's station, whatever that may be, for as soon as arrogance sets in, Mother Nature, competition, and customers will conspire to bring one down. The ecosystem is remorseless and is always looking for subjects who need to learn lessons and become more balanced.

Shoulder of lamb, creamed kale, cranberry beans with lamb bacon, baked apples, and polenta welcome the fall season. Now, that is excellence! Humbly striving for excellence, with your help.

Virtue of Travel
November 16, 2017

How does one travel, while living on a farm?

When the question is posed, "Why don't you go see the world?", our response is: "We are traveling in our own world, more deeply than ever realized or anticipated".

First of all, we log a lot of miles in our truck, endlessly traveling from Pike County to Hamilton County and back. About 300 miles a week or 15,000 per year, which amounts to respectable distance traveled, by most standards. And we make a few trips to Ontario, which adds another 2,000 miles. AAA thinks we are rather worldly.

When it is implied we might broaden our horizons periodically, I point out that our horizons are as magnificent and glorious as any. What would one do with more glory in one's life?

Over the past ten years, we have engaged in an astounding adventure, that provokes us entirely, through body, mind, and spirit. The challenge of stimulating an ecosystem in Pike County to provide nurture for people, known and unknown, 80 miles away, propels one into a journey rich with stimulus, learning, color, relationships, and immense satisfaction. And there is no charge for such adventure, other than giving oneself to it entirely. The interesting people one meets when traveling are matched, in our world, by the animals we raise and customers we have come to know, both full of rewarding personality. The art museums we are missing are matched by changing seasons of nature, which drench us daily with fresh beauty. And we incur no jet-lag in our travels, except for rising at 5 a.m. on Sundays.

An Erudite Conversationalist   *Mt. Magnificent*

The local and delicious food one encounters, when navigating foreign cultures, is rivaled by the culinary journey right at our own table. Susan's Soulful Kitchen regularly takes us to Korea, Morocco, Vietnam, India, France, Italy, and Mexico. Remarkable by nearly anyone's standards.

A lot of traveling is realized simply by maintaining a rhythm in life, allowing one to navigate deeply into experience. The constant back and forth of rhythm is like the sawing of wood. Eventually a point of release is realized, and new reality dawns. Our entrenched rhythm, that keeps us feeding animals every day, cooking food every week, and attending markets every weekend, is thankfully always teaching and liberating us, despite its confines.

A wise man can travel in an armchair, by listening to that which is without and within. It is the listening that is so difficult, something we are not taught in schools. If one can listen, one can move to new horizons quickly. The "ugly American" covers a lot of ground abroad, but doesn't bring much back with him, because of noise within, obstructing his ability to receive.

In our own unrelenting busy-ness, we are invited to listen to: employees, animals, equipment, weather, food-on-the-stove, and customers. Such listening expands us, keeping our travels vital and rich.

So, when the observation is offered that we don't travel much, I respond, but we do, beyond our wildest dreams.

Three species on one acre together – 140 bovines, 100 laying hens, and 1 guard dog. That is a sight one would have to travel far and wide to witness elsewhere.

Pork chops, kale, and sautéed pears. We have an ancient pear tree on the farm that produces hard, mealy fruit. Susan magically transformed a batch of these unruly pears into a true delicacy, by sautéing them with wine, butter, and some sugar.

May we always travel toward our dreams.

Language & Grassfed Meats
December 7, 2017

The most important ingredient to grassfed meats is not a beautiful shepherdess or great green grass, though invaluable, but carefully chosen words.

Words and language are links in chains that bind together many facets of our business. These include: building infrastructure, managing soil and plants, rearing livestock, maintaining equipment, shipping livestock, having livestock processed, transporting meat, cooking meat, marketing it, and delivering to customers. Those are nine distinct activities to be managed. If performance within one is off, impact ripples throughout the line of connection. When things go awry, it is usually because of communication; someone said one thing and meant another, generating degrees of confusion and unintended outcomes.

Too often the source of poorly chosen words is the self, at least in my case! Doesn't the following statement, "Move the cows to high-ground only when it floods," have a different meaning than, "Move only the cows to high-ground when it floods"? In the second, the sheep are left behind to swim. So, the manager inquires why the sheep weren't moved during the flood. The answer is simple, "because you said move only the cows." But that isn't what I meant! But it is what I said. All because a word was unintentionally moved from one location in the sentence to another.

I recently asked the meat processor to make some lamb patties for us, assuming they would be just like the beef patties we always request—three round, 5-ounce patties to a package. Well, the lamb patties came back as four square, 4-ounce patties to a package. I also asked them to make lamb bacon for us, assuming they would make it just like our pork bacon, cut thickly. The lamb bacon came back cut thinly. Both of my assumptions failed to communicate effectively.

With enough misplaced words and assumptions, chaos finds an opening. Adding to the challenge of communication is the fact the English language is always evolving. Stately nouns are disrespected by creative speakers and turned into unconventional verbs. For example, it used to be considered common practise to arrange food on a plate. Now one simply plates food. What happened to arranging food? In addition, an annotated and abbreviated new vocabulary, created by the surging digital world, confuses matters further. "Organic" used to apply primarily to rich, black soil; now it has been co-opted to determine the kind of advertising Facebook will support. Further, language is now being extended with emoji symbols—a new word itself, bringing incremental form to the English language. If you are happy about saying something, why not just say so, rather than include a stylized picture of a smiling face? For those who are more visual than literal, perhaps the emoji symbols are a welcome addition to the English language. Apparently, the same word in Chinese can carry three very different meanings, depending on "tone." At least we don't have to deal with nuance of tone, or do we? We have all been told by parent or spouse, "I don't like your tone."

So, communication is complicated. When it is executed well, great products result, like poetry, beautiful buildings, and grassfed meats! When done poorly, mediocrity is compounded and even harm can be imposed.

One of our neighbors, Nancy Stranahan, is an excellent communicator, who chooses her words very carefully, to profound effect. She is the Director of Highland Nature Sanctuary, which is creating a series of nature preserves throughout southern Ohio, referred to as the Arc of Appalachia. If you want to read great writing about important topics, take a look at this winter's newsletter, especially the first article addressing the *"Irrepressible Resurgence of Life"*.

 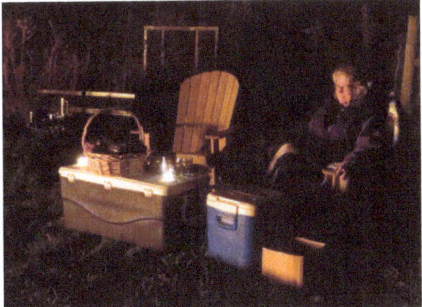

Susan and I enjoyed a potent evening last Friday, before both the super-moon and a powerful bonfire, extinguishing a year's worth of brush. We were at a loss of words before this display of power and light, as are guard dogs when eating breakfast. Wordlessness can be sacred.

One of our favorite meals is Shortrib Burgers. We often enjoy them with organic French fries roasted with beef fat, but here they rose to an even higher level accompanied by potato pancakes and baked apples.

May carefully chosen words enhance the well-being of us all.

Breeding Season
December 14, 2017

We moved bulls in with cows this past week.

Three bulls had to give up their idle ways to accompany the ladies for 45 days of breeding. They were easy to move and ready to comply, once reunited with the matriarchs. Bulls are delicate to manage, because if they start strutting and fighting, nothing moves them anywhere. But ours are docile, and despite some fighting, they do comply, with patience on our end.

We have five bulls now. One is a yearling, who just arrived last week. We will not use him until next year, but wanted to give him time to acclimate. We learned the hard way years ago that turning a new mating male in with a group of mating males is recipe for gang warfare. We turned a new mature ram in with a group of five young rams the day breeding started. The five rams, who were familiar with each other, ganged up on the new one and killed him. What a shock that was. We have learned, and are being cautious with our new bull.

We have three bulls in with 75 cows and one held back in reserve, in the event the first string needs relief. It is cheap insurance to have extra bulls. Nothing is more important than a narrow and effective breeding season, making management of the herd the rest of the year so much easier.

We had a visitor from Colorado last week. Clark spent a few days at the farm helping to move bulls, load hogs, repair fence, and change tires—the kind of visitor you wish would drop by once a week or so. Pictured above, we have just moved the bulls in with cows, and have given the cows a new break of grass.

Five of our newsletters are featured in the current Winter edition of *Edible Ohio*, as examples of telling the story. We have hard copies of the edition to give out and will have a link to share in January.

Edible Ohio does a magnificent job of telling the story of our foodshed. The team there consists of positive and uplifting people, who are helping to weave the essential fabric of our community. May their good work continue, with our support.

May the breeding season go well for bovines and the holiday season for people. Would that it were as simple for the latter as it is for the former.

Water Worries
January 4, 2018

Cold weather challenges flow of water! We are all dealing with this, whether frozen pipes in the basement, burst pipes on the side of the house, or other symptoms mysteriously affecting the flow of the world's most precious resource. It is one thing when one runs a household of two or four or six. Alternative solutions can be improvised for a while. It is another when one manages 500 animals who need water on a daily basis, and no other source is available. In winter, livestock's demand for water is less than in the heat of summer, so one has a grace period of about 24 hours to fix problems.

Our water tubs in winter are recharged from the bottom, with flow being arrested by the yellow float at the top. On a daily basis, in freezing weather, we clear ice from the surface so animals can drink and ice does not play with the float. Ice will stretch the string between float and shut-off valve, allowing water to overflow the tub. This fairly quickly shuts down the pump in the spring-house, preventing refill of other tubs on the farm.

So, when Kathy reported she went to break ice on this tub and found it to be half empty beneath the surface, it was clear a mysterious problem had arisen. Water was leaking somewhere!

The first step usually employed is to ignore the problem, hoping gods will deliver an unexplained solution by morning. Since one has 24 hours to resolve the matter, and when hour 20 arrives without divine intervention and the forecast is for continuous sub-zero weather, a sense of urgency sets in! I thus braced myself to attack the matter. The temperature had risen to about 20 degrees, so I had a nice, balmy, momentary opening in which to play with water, wrenches, fittings, and ice.

The first step would be to remove as much ice as possible from the tub, in order to eventually turn it over and examine what was transpiring beneath. The next step would be to drain the tub of water, which typically requires unscrewing the plastic plug at the bottom. Well, the plastic plug was solidly frozen in place, and was totally unresponsive to efforts with vice-grips to turn it loose. What to do? Go have lunch, of course.

So, over lunch, I shared with my good wife the current predicament. She, having studied plumbing in both culinary and law schools, immediately suggested taking a thermos of hot water and pouring it over the plug. This was an embarrassingly insightful suggestion to the acting plumber, which worked perfectly. Three hundred gallons of water poured out, creating a sloppy mess, that was turning to ice with every increment of procrastination.

The next step was to turn the tub over, which is not much of a problem when no ice is attached to the inside of the tub. When ice is attached, it is quite difficult to lift it up-and-over in one motion. But I was able to lift one end of the tub enough to place the empty two-gallon thermos-jug under it. I then squatted down, and push-hoisted the ice-laden tub up-and-over. Making progress…

Next step was to empty the 55-gallon barrel beneath the tub, housing plumbing fittings, that had filled with water. One can empty three quarters of it with a 5-gallon bucket from a kneeling position, but the last increment requires lying down on the cement pad. I scraped and shoveled as much water off the pad as possible, and laid down some towels brought just for this purpose, so clothing wouldn't get wet. The barrel was thus emptied, only to immediately refill with ground water from the bottom, some of which came from the 300 gallons that had just been dumped out! The barrel was thus patiently emptied again, with gloves getting wet, despite care taken. I would periodically retreat to the truck to warm hands and had also brought a second set of dry gloves, having learned from previous encounters.

Finally, the barrel beneath was empty of water, and one could try to identify why and where the leak was transpiring. The two hose-clamps at either end of the hose had not come loose. The next thing to check was the plastic female socket that receives the male end of the hose, and activates flow of water when plugged into. That female piece screws into a ¾-inch, threaded elbow with a compression fitting on one end, connecting to one-inch black pipe, buried three feet deep. Tightening the female socket is easy, requiring a pipe wrench and about ten seconds of time. If that doesn't fix the leak, then the problem is in the elbow and compression fitting, which are buried in mud and are about six inches deeper at the bottom of the barrel. That repair takes about an hour of working upside down in a very tight space, and is not much fun. I was rooting for the female socket.

The problem proved to be the loosened female socket, to the plumber's great relief. Once the socket was tightened and hose plugged back in, the tub was repositioned and refilled. Despite precautions over the previous two hours, I was somewhat wet and beginning to feel cold. But it looked like we were on the downhill side of the problem.

However, while the tub sat empty, residual water in the valve at the bottom of the tub had surreptitiously frozen. So, once the tub refilled, the float wouldn't entirely shut off, and water began running over the side, creating more mess. Well, desperate problems call for desperate measures. I took off my glove, took out my pocket knife, opened the blade, plunged my arm into the cold water, and was able to find the spot on the valve

where a slight build-up of ice was preventing it from closing. I scraped enough ice off that the float took hold, and flow of water was arrested.

In tepid triumph, I climbed into the warm truck, turned the heat to full force and drove back to the house, feeling increasingly cold, and very relieved.

We have noticed in this cold weather that cows come to drink in groups rather than individually. When ten cows are grouped around a water tub, each drinking five to ten gallons at a time, followed by another group of ten, followed by another, it is difficult for the 1 hp pump in the spring-house to keep up with demand. When this happens, it loses prime, and shuts off.

Last year we hooked up to the county water-system, to maintain as a backup for spring water. County water delivers lots of volume and pressure, and has served us well over the past 12 months as a backup, but comes with a monthly bill.

Yesterday, two days after the first incident described, we noticed two water tubs were nearly completely drained, and the spring pump was off. Given that two inches of water in the bottom of a tub in freezing weather is going to freeze the valves, it was important to refill those tubs quickly. So, I turned on the county water, feeling glad we had invested in that insurance. But nothing came, no matter how long I waited. Frustrated and perplexed, with dark approaching and cows wanting water, I was able to re-prime the spring pump, and over two hours slowly refill those two tanks, without losing water pressure.

That evening, in reflection, I realized the county water had probably frozen at the water tap and meter, out by the road. So, the next morning, I filled a ten-gallon thermos with hot water, poured it into the barrel of the tap, and behold, watched the meter start to run again. Now a bale of straw sits atop that water meter, and our insurance policy is back in effect.

I apologize for this long account of our challenges with water during the past week. Each of the details recounted, however, loomed very large when trying to solve essential problems. This is what our week has been about. Ultimately, it has been about stewardship of livestock. We do all we can to give them the best life possible, especially in extreme conditions, so we can deliver to you the highest quality food possible.

A final note on this subject. After working three hours on the water tub, two days ago, I began to notice, toward the end, a little man appearing above my head. He was carrying something that looked like a pitchfork, and he had a sly, unseemly grin on his face. He began taunting me, asking if I didn't want to go back to that comfortable office job I once had and how did I like being miserable in the cold. I tried ignoring him for a good while, but he wouldn't take the hint. Finally, I lost my cool, and grabbed the frozen pipe wrench, lashed out, and struck him down to the ground. Then I maliciously poured a bucket of cold water on top of him and stomped him into the snow, making myself feel a lot better. He seemed to disappear, but I yelled at him anyway, never to come back! Our demons are too often close by, but they can be overcome if we remember who we are.

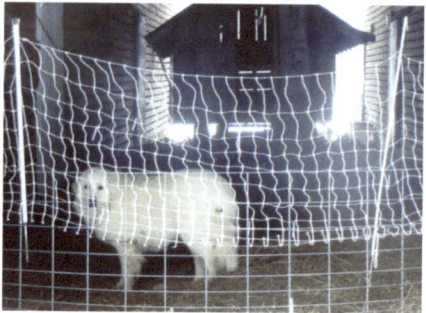

Hogs happily pile up, like extended family, to stay warm. Laying hens and Coquie have moved under second roof, closer to warm water.

Spatchcocked chicken, that eluded the last newsletter, served with mashed potatoes, brussel sprouts, and a tomato salad. Shortrib burger, potato pancakes, and baked apples. Potato pancakes came from left-over mashed potatoes.

A rediscovered favorite – brined pork chops. The chops are brined in a gallon or so of water with a quarter cup of sugar, quarter cup of salt, and herbs of interest, for three hours or longer. They are then seared on a hot grill or frying pan for two to three minutes per side and placed in a 400 degree oven for 5 minutes. Juicy, succulent, and flavorful! So good, it is hard to eat only one. We are really liking Berkshire meat, and think specific breed does make a difference in quality of pork.

The accompanying sweet potatoes were hand-delivered by a farmer and friend, from his harvest in North Carolina.

A number of well-informed people are referring us to a strange, mystical practice called "rest." We have heard about this, but are suspicious of New Age fads, especially one that doesn't require motion. Nevertheless, we have agreed to give it a try, before dismissing it entirely. Accordingly, capable colleagues, Beth & Bob Gehres, will be standing in for us at the Hyde Park Market this Sunday.

As the New Year rolls in, let us resolve together to keep water flowing, demons at bay, and good food coming!

Fearless Cook
January 16, 2018

Georgia Gilmore presented resistance through her food. She would cook for gatherings of the civil rights movement during the late 1950s, donating proceeds back to the movement, as it resisted segregated busing. She was so successful at raising money for the cause through cooking that others began emulating her model. This was all done within the African American community, out of sight of authorities, and was referred to as "money from nowhere." But she was a fearless woman who was willing to testify in court, and was willing to lose her job at the National Lunch Company as a result, in pursuit of justice. She would not back down from tyranny, and subsequent to losing her job, opened a restaurant in her house to continue the quest for freedom of her people.

Over the years, we have explored what goes on around the dining room or kitchen table. The list is wonderfully long, and includes:

1. newspapers read,
2. homework conducted,
3. groceries stacked,
4. food presented,
5. nutrition administered,
6. meals savored,
7. stories recounted,
8. conversation exchanged,
9. community gathered,
10. values taught,
11. emotions shared,
12. concepts pondered,
13. plans made,
14. babies conceived,
15. babies born,
16. rest taken,
17. travel realized.

And now we have another activity for the table, thanks to Georgia Gilmore:

#18. resistance presented.

As we ponder the meaning of Martin Luther King's words and actions, whose speech "I Have A Dream" is among the most inspired in the English language, what might we resist, as we prepare food and bring it to the table? We might resist the concept of selecting members of our society based on pedigree and genetics, as if they were strains of seed corn. The inscription on the Statue of Liberty proclaims:

Give me your tired, your poor,
Your huddled masses yearning to breathe free,
The wretched refuse of your teeming shore.
Send these, the homeless, tempest-tossed to me...

It is not "PhDs" from one corner of the world who have made our country great; it is wretched masses yearning to breathe free. It is those with fire in their bellies, who expect nothing and are willing to do anything. They understand dignity of labor, and are humble about opportunities before them. They do not arrive on our shores laden with entitlement. Their immense character cannot be determined by race or country, but only by terrible circumstance.

Let us resist the notion of racial purity, for it does not exist in nature and will never withstand the tides of time. But during moments, it can prevail and inflict much damage. So, by example of Georgia Gilmore, we are now called to summon courage and stand up against the rising cruel refrain.

Ewes paw for grass.

The sump pump for the drain in our basement kitchen froze yesterday. So, we are building an igloo by which to heat it up. It has been five hours since construction, and no action yet, but I put my arm in there and it was toasty. Anybody who guesses how long it will take to thaw out, receives a free prepared food of choice from us.

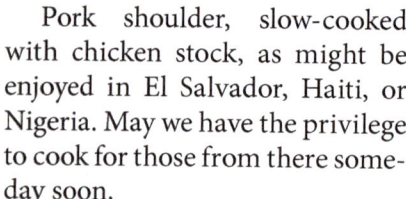

Pork shoulder, slow-cooked with chicken stock, as might be enjoyed in El Salvador, Haiti, or Nigeria. May we have the privilege to cook for those from there someday soon.

In tribute to fearless cooks.

Last Snow
February 6, 2018

As last snow departs, we both miss the beauty and welcome the change.

Feeding hay on snow-covered ground works well for plants, livestock, and machinery. Snow provides insulation to plants, keeping fescue and clover protected from freezing temperatures. As long as ruminants receive hay and water, they are quite content in cold weather. Machinery rides well on hard surfaces and the ground is not torn by heavy loads. The landscape also takes on an unusually quiet, serene, and inspiring tone when covered with snow, very distinct from other times of the year. As we seem to be realizing less snow than in decades past, interludes of mid-winter serenity are increasingly precious.

On Tuesday morning, Kathy and I brought the cow herd in and separated out the bulls. Their 45-day tour of duty had been completed, and we will find out in April with what degree of efficacy. Now they head back to solitary confinement, to be amused only by each other, without the charms of the fairer sex to distract and entertain. We strive for a narrow calving window, so all calves are about the same age, requiring about the same management. We don't want calves who are two-days old in the same herd with calves six months old. Once calving has been completed, we resume moving the herd quickly around the farm, and don't want to leave two-day old calves behind inadvertently. So, the breeding window is brief or narrow. Cows have two chances to be bred during that time, and it they don't conceive, then they are culled for other purposes. Building a good cow herd is a long and slow process.

Working in winter conditions builds appetites. This one was recently rewarded by a repast of: ribeye steak, hand-picked Carolina rice, and mushrooms & green peas. So good and so satisfying.

The answer to the riddle about the defrosting sump pump is it took about 48 hours, at best guess. Part of the issue was operator error, in that the heater was turned on low rather than high. Therefore, anybody who ventured a guess is a winner, and we look forward to including your prize with your next purchase.

In the flowing melt.

Rain Blast
February 26, 2018

After two inches of rain fell onto saturated ground Wednesday night, these cattle felt stressed. When they are bunched-up like this, they are registering discomfort. We have received so much rain over the past two weeks, that we have had to change strategies on managing cattle. We had been unrolling hay behind the cows, so as not to drive equipment onto ungrazed pasture. But animal impact on grazed, saturated pastures is rendering them too muddy to support a tractor loaded with hay. So, we are now unrolling hay in front of cows. We have to drive over new feed, which is not optimal, but hay is distributed on clean grass, and impact from the tractor is minimized due to untrampled sod.

Cows are much more relaxed once they have access to new grass and fresh hay. They and we feel a lot better.

Just to keep days even more interesting, the five-year-old compressor for one of our large freezers decided to die late last week. Industrial compressors are supposed to last 15 years, but ours opted-out just after the warranty expired. We therefore had a walk-in freezer full of meat, registering ever-rising temperatures. After staring at the situation in disbelief for a few hours, I finally conceded the freezer had to be emptied. I thus shuffled 70 boxes from the malfunctioning freezer to the one still working, and stacked them wherever possible, including in the aisle. Fortunately, we have a second freezer and there was room for stacking. The new compressor will be installed Monday morning, and then I have the pleasure of moving all those boxes back into place!

Susan and I regularly employ several of our prepared foods as part of our weekly food ritual. Shortrib burgers are a constant presence in the routine, served on an open faced, Blue Oven English Muffin, with homemade mayonnaise. Organic French Fries are roasted with some rendered beef fat to accompany, becoming irresistible. Bolognese Sauce is also a weekly stalwart, that provides copious leftovers. We always look forward to both of those meals. To be recommended!

See you on Sunday, despite predicted blasts.

Side by Side
March 31, 2018

Landis Weaver and I worked side-by-side to construct this sorting corral. It took several afternoons to do so, and then we tested it out. A number of remarkable aspects to the experience surfaced. First, the metal panels employed (made by Priefert, out of Texas) come with chains, high and low, at one end, which hook to slots, high and low, at the other end of the next panel. Thus, hooking panels together works readily, creating a solid structure. It is gratifying how quickly one makes progress, chaining one 15-foot metal panel to another.

Large, sturdy gates with bowed frames, are then placed between panels, wherever desired, enabling multiple designs and changes in design. This flexibility is so much better than the system we employed before, installed 30 years ago, which incorporated pressure-treated wooden posts, sunk into the ground, and attached together with nailed-in oak planks. Once those posts were sunk, better ideas were not allowed. No changes could be made to design. Whereas the Priefert system is infinitely flexible and malleable, as gates and panels can be moved at any time.

Testing the design went better than anticipated. We brought a group of 30 dairy cows through, who needed new eartags. The flow of people and animals went smoothly. We were able to accomplish the task with minimal impediment, to our surprise.

The best aspect of this process was working with Landis, side-by-side. We did everything together—carried, chained, designed, redesigned, grunted, sweated, swore (me not him), got tired, exclaimed, experimented, handed tools back and forth, anticipated each other, finding wordless flow that was precious. Susan asked what we talked about. Well, which end of the panel to put where. "That was all?", she implored. "Yes, that was all we needed." While, at the same time, we were communicating fully: improvising choreography, so we worked with rather than against; always regarding the other in generating the next step; moving fluidly, in a dance of two; eyes, hands, bodies, and grunts continually interchanging to convey messages, in ready improvisation, over the course of six hours.

There was a marvel to it all – an Englishman and Mennonite, 30 years apart in age, working seamlessly together. The experience was crowned at the end, when Landis, a man of few words and impassive facial expression, stood back, paused, raised his great, green eyes, cracked a big, beautiful smile, and gratefully remarked "This will do."

A farmers market dinner: Elmwood's kale, Eaton's sweet potato, Backyard Orchard's apples, and Grassroots' pork chop. Such fare is some of the best insurance against poor health money can buy.

We have just migrated to a new website, incorporating shopping into the body of the website. This change has been a major undertaking. It has been like laying a whole new water system, with an infinite number of new plumbing fixtures and valves. I am having to learn where each is located and how it functions, which is somewhat mind-numbing and anxiety-provoking. But such appears to be all part of the journey in this digital age. You will have to sign up again with a password to make purchases, for which I apologize. During this transition, we will have some challenges, no doubt on this end, but please bear with us and help by communicating any issues you see. We hope this considerable effort and investment is a step forward. Let us know what you think.

In this powerful moment of the year, which celebrates ancient traditions of liberation and redemption, let us continue to work side-by-side, among customers and farmers, among rural and urban, among neighbors, among religions, among all backgrounds, so we may stand back and gratefully remark, "This will do."

Spirit Quest
April 21, 2018

These ewes are about to engage in a great spiritual event – giving birth.

Animals, unlike humans, do not seek spirituality. They own it, are part of it, even create it. Giving birth is an extraordinary experience, and it is in the extraordinary that the spirit lies. A mother delivering a child is clearly in an exceptional dimension, which feels sacred to most witnesses. As these ewes bring forth new life over the next month, we honor them, and pay homage to them, for the great miracle they propagate.

It feels like spirituality is inherent in animals, whereas humans have to seek it. The seeking is a difficult but wondrous proposition. Some do so formally through organized religion. Others do so informally, but vividly, through values lived. The quest for claiming greater context to daily actions seems to be essential to the human experience. This quest knows no boundaries or definitions.

Several weeks ago, this member of the Grassroots team gave a brief presentation, entitled *Spirit Quest*, over lunch to the City Club in downtown Cincinnati. *Spirit Quest* is the story of one man's journey to claim context for his purpose, summarizing work on our farm over decades. This includes developing: a grass-based dairy 30 years ago, a wetland-mitigation-bank 20 years ago, and a production & marketing system for grassfed meats 10 years ago. This journey could never be highly rationalized at the outset, for which a price was paid financially and emotionally. Rather it was primarily an intuitive process, buttressed by rationale in part but, more importantly, by "calling." When calling and rationale combine forces, magic can arise. We have experienced such magic over time, and are so grateful to be recovering early losses as we move into a sense of abundance.

Yet the rigor of the journey never ceases, as the spiritual path is perilous and is not for the faint-of-heart! Free falls and periodic tumult seem to be inherent in its design. We experience that turbulence and currently are, but with ongoing faith and work, equilibrium always seems to reestablish itself, in its own unpredictable way.

In sum, a spirit quest calls for far more than one has to give, but rewards one with far more than ever expected. This enables one to conclude that life is well-lived, which might be as good as it gets.

At conclusion of the presentation to the City Club, one member commented that such a spirit quest is not something one can engage in as a youth, as it is not really available until later in life. That observation feels accurate, though there are always exceptions.

Last weekend, we gave a tour of our wetlands to a group from Arc of Appalachia. Arc of Appalachia (aka Highlands Nature Sanctuary) is a partner of ours, in that it oversees the conservation easement placed on our wetlands. These visitors were so supportive and interested in what we are doing, that they left behind more positive energy than they took. That is abundance in itself.

In several weeks, we are looking forward to hosting a class of three year olds, enrolled in the local Head Start program, further enhancing spirit.

A recently cooked, spatchcocked chicken. These pasture-raised chickens are markedly more flavorful and firmer than conventional ones, aside from being more nutritious. Spatchcocking is easy if you have a heavy pair of kitchen scissors or poultry shears. One cuts out the backbone, butterflies the carcass, and cooks it at 400 degrees for 40–45 minutes. Brining beforehand adds succulence. This has become Susan's preferred method for cooking a whole chicken.

May we each know a spirit quest, when the time is right.

House of Love
July 7, 2018

*This 150-year-old soul is being rejuvenated to
provide house and home for another century or two.*

It was fifty years ago that we first envisioned living in this hand-crafted house; five years ago we developed detailed plans for its renovation; two years ago we began deconstruction; and three weeks ago reconstruction began in earnest.

This house was built just after the Civil War in 1867 and sits on a hilltop with a stunning view of a sweeping valley. Upon deconstructing several years ago, we discovered Chestnut beams throughout, which no doubt were logged in the nearby forest. Few nails were employed in construction, beams were anchored together through mortise-and-tenon technique. Some of the beams are 20 feet long and some planks of Poplar 18 inches wide. Such dimensions for lumber are not available today, and American Chestnut is now extinct. Ceilings are nine feet tall and windows six feet.

This house stands with a grace and perseverance that speaks of the ages. Its classic dimension endures, as the classic always does.

In developing plans to renovate, the tall dimension of the structure seemed to ask for wings to create balance. We obliged with a porch on one side and an expanded kitchen and dining room on the other. A masterly bedroom was not obvious, until it appeared beneath a raised roof, with breathtaking regard for the valley below. An office under metal roof will now sit in the unfilled corner below, with full view of farming activities.

The more we contemplated plans for this project, the louder they spoke. We lived with them for five years, during which their clarity and beauty became evermore compelling. They ripened, awaiting our courage to implement. We have finally summoned such to do so.

This is an immense project financially, emotionally, and logistically. It will call on everything we have to bring it to fruition. The financial advisor sees simpler solutions to our challenge, like tearing down the building and constructing something for half the cost! The spiritual advisor celebrates listening to the forces at play offering a complex response.

We are responding to circling and overlapping messages in reaching this conclusion:

- The house is tired, and needs to be either rebuilt or torn down;
- We are tired, and need to expand our housing stock to properly invite successors forward;
- We want to make our current home available to our successors;
- Good housing is essential to attracting good people;
- This house has stood prominently for 150 years and, in deference to our hard-working predecessors, we want to support it for another 150;
- Chestnut wood is extinct but will be carefully preserved and on view in this house;
- The right builder is fortuitously available to bring this project to the fore;
- We want a home big enough to welcome and accommodate friends and family, who travel long distances to see us;
- We want a home with wings on it, so we can fly wherever we want, at no cost;
- We want a home that inspires us every day, as we move into older age;
- We want a home that fully dignifies the magnificent land upon which it stands;
- In the end, we want a home that speaks of love for the land by which we live, the natural life that thrives upon it, the exquisite food that comes from it, the exceptional employees and partners who support it and us, and the invaluable customers who ultimately give us reason for being.

We want to invest in love. And as we all know, doing so is risky, given that love always demands more than we have to offer. But Susan and I are intimate with investing until it hurts, as that is the way we know to live. So, we have made the non-rational, but fully informed and inspired decision, to create a shrine to *the loving process.* This investment is a lot more compelling to us than investing in the sterile stock market. Over time, we think our returns will be considerably better than 7% per year, garnered by the latter.

But the most important reason for renovating this house is finding an excuse to incorporate these two stones into its design. They are seven feet long, by 28 inches wide, by 8 inches thick. They surrounded the opening to a cistern, which we filled, and will find new quarters as raised hearth-stones in both the kitchen and living room. We invite you to come and sit upon them with us, before the fire, in golden years ahead!

We look forward to seeing you this Sunday at Hyde Park, on Wednesday at Blue Ash, and on Thursday at Bexley. We will not be grilling sliders this Sunday, as our grilling team will be lakeside. Beth & Bob will be manning the market on Sunday, while Susan and I are creekside in Pike County.

You are essential partners with us in building a house of love for the next 150 years, which is no small feat!

In humility before that house.

*Herb garden in full expression,
for our prepared foods.*

Moving Dirt
August 24, 2018

Creating an authentic life requires moving dirt!

If you are concerned you have not yet moved enough dirt for your life to crystallize as you would like, fear not, for we have just moved 200 truck-loads (3,000 tons) from this creek-bank to where our new driveway intersects with the state road—in quest of authentic living, and in quest of safety, so we aren't run over by unseen logging trucks!

This has proven to be quite a project, with benefits and costs at both ends. The creek was channelized probably 80 years ago to optimize production of grain crops. One result is creek banks that are now 8–10 feet deep and steep. These are dangerous to man, machine, and beast, and are susceptible to erosion. They are slowly but steadily collapsing into the stream, with soil being swept downstream.

We have been wanting to address this issue for a long time, but given that we have nearly five miles of stream banks, it has been somewhat paralyzing to discern where to start, and what to do with all the dirt and then, of course, how to pay for it. And how important a priority is it? The previous generation left the problem to us. Do we do the same, postponing action as long as possible?

Given that we are discernably aging into our mid-60s, we realize we had better get a handle on this authentic-living-thing pretty quickly, lest it slip through our fingertips entirely. So, we incorporated moving-dirt into the budget for our new house.

Five hundred feet of creek bank is now being peeled back and grassed-down, which will both stabilize the bank and reduce flooding of our bottoms. The sloping banks will eventually be planted with willow trees to provide shade for aquatic life.

That dirt has been transferred to our front drive, so vehicles will intersect with the road at a level plane, rather than a dangerously steep one. We figure this to be an investment in personal longevity.

So, now that part of the stream bank is at rest and we have figured out how to enter and exit the drive of the new house without mishap, how about the authentic-living-thing? Well, here is a brief account of 48 hours this past week, which

were stimulating, and though tiring, offered testimony that moving all that dirt just might be working after all.

Early one morning, Bo and I brought our diminishing flock of yearling harvest-lambs to the barn and sorted four of them, two of which were to be brown, for Yousef and three of his companions. They are Moroccan and were celebrating the Islam holiday of Ɛiyd. Ɛiyd recounts God testing Abraham's faith by asking him, if he is a true believer, to sacrifice his only son. As Abraham raises his knife to do so, God intervenes, and offers a lamb instead. Yousef recounted this belief, and then he and his companions proceeded to harvest four live lambs on site, in ceremonial fashion. The next morning, Ravshan, who is from Ubekistan, showed up for two more lambs, for the same reason. These are interesting people, who know the quality of our lamb, and are willing to drive two hours to pay its

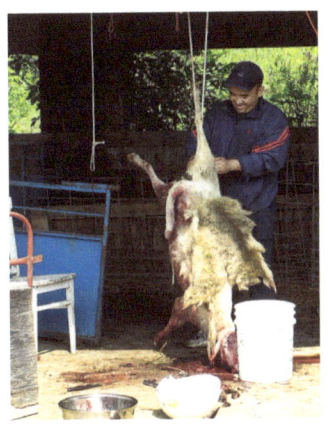

premium, without negotiation. How did that phenomenon unfold? Only over time and with persistence.

And then Bo and I fed the animals, which is always a pleasure, as long as everything goes right, which it did. As the cows are no longer lactating, they are gaining weight in knee-deep forage. Hogs are enjoying freedom, romping around in their three-acre paddock. Hens actively eat grass and grain to produce thick yellow yolks. Roosters, who were supposed to be hens, live the-life-of-Riley, crowing about everything and producing nothing. Ewes are in deep grass, lactating well and raising clean, hearty lambs. Our neighbors, the Zimmermans, harvested beautiful Rock-Cornish-cross meat birds for us, that are notably clean and well-managed, which Bo and I packaged and labeled into the evening. We weaned the last of yearlings and moved them to their own pasture. I conferred extensively with the contractor moving dirt and the builder renovating our old house.

We have moved a lot of dirt to arrive at this point. One has to sculpt, deny, fail, imagine, persist, and create one's own structure to claim the authentic life within, that begs to be released without. Hopefully, you won't have to move 200 truck-loads to do so, but if you do, know that you are kindred spirits with us and we welcome your company out here in Pike County at any time.

This evening we were blessed with two culinary treats, in supplement to acute delight offered by fresh summer produce. Susan created a goat's-cheese-and-leek quiche, that was addictive. Further, and for the first time, she cooked chicken breasts by immersing them sealed and sous-vide in a tub of water for 2 hours, at 140 degrees. She then seared them quickly in olive oil in a frying pan. The result was the most moist and tender chicken breast we have ever savored. Both were extraordinary.

I am somewhat embarrassed to confess I love my wife an increment more after every one of these meals that we have. But why not? Such is the virtue of great food.

In tribute to the power of dirt to foster authentic living.

Rainbow of Life
November 2, 2018

Rainbows appear when we need them.

In the midst of cold rain, grey skies, concern over inventory, and wonder about transitions, a rainbow graced our farm this morning such as we have never before witnessed. It was a full rainbow visible from one end to the other, with a great arc and a complete, thick spectrum of colors. It was magnificent, breathtaking, and instantly inspiring. It filled the heart with vitality and a sense of purpose, thus warding off the day's woes, enabling us to continue this great and demanding journey.

One of the privileges of life on the farm is anecdotes of inspiration, such as this, which surface unexpectedly. It never seems to fail that when the footstep or heart are heavy, something shows itself to dispel the burden, if the eyes are open. What a gift it is to receive these showings, like dividends upon investment that appear out of nowhere.

We have a number of barn cats on the farm, one of which is Coffee. They seem to fit in so naturally. Coffee lives near our apartment, where visitors now stay, keeping it clear of mice. We feed her in the corncrib, so dogs don't avail themselves of her fare. It always feels good to feed Coffee. She improves the day, in her silent dignified way. Her mate, Tea, is mostly white and lives at Landis' house. An itinerant grey cousin lives at our other barn. Each fills a void, providing color in the rainbow of animals on the farm.

As mentioned months ago, we have started taking down dead ash trees on the edges of pastures, in preemptive fashion, so we aren't faced with a crisis when they fall at inopportune time into pastures and creeks and upon fences. This has been a fairly successful strategy for avoiding crises, but we have an unexpected outcome to deal with. That is what to do with all of the tops now stacked along the sides of the creek. With flooding, those tops will sooner or later end up in the creek, and then we will have a closer look at crisis management. So, we can leave them where they are and hope they decompose before too much flooding occurs, bring in a chipper and convert all of that biomass into wood chips to provide bedding, burn them to accelerate decomposition, or haul the tops to higher ground where they can decompose without risk. We will probably try some of each, and will keep you posted.

It is disconcerting to observe how much of the forest is composed of ash trees, now dying or dead. Along creeks and edges of forests, ash amounts to probably 90% of species. Deeper in the forest it is closer to 25%. Nature will fill the void, but only over time. And what species will be next to fall to invasive insects?

These lamb chops were some of the best we have ever had. The lambs were 18 months old, so the chops had more fat than usual yet were tender as tuna. They looked as good as any sushi and, barely cooked, made our eyes water in gratitude.

We look forward to seeing you this week, as we move indoors to Clark Montessori, 3030 Erie Avenue. If you care about nutritious food, please continue to support vendors at the market. We have bills to pay year around, and are dependent upon customers to keep us vital, as you are dependent on us to keep you vital, which is a beautiful and balanced connection.

May the rainbows of life continue to inspire us all!

Coyote Blossom
November 29, 2018

This past Sunday morning, a beautiful coyote sat where this white post stands and stared at me intently. I was putting up the hot-wire and posts in the background, when a distinct presence appeared over one shoulder. I turned to experience a sensation never before felt—a coyote calmly sitting 30 yards away with ears perched observing my every move. Once I caught my breath, I stood and watched, and she watched back, so I watched some more, and she kept doing the same. After a long period of mutual regard, I felt compelled to move on with my work, but did so slowly so as not to startle her. She wasn't startled at all, but just kept staring at me. Her purpose began to seem far greater than mine. So, I finally put down the reel and posts in hand, and squatted to her eye-level to resume our communication.

Upon quieting myself, I slowly began to hear the message she was offering. And then it became magnificently and movingly obvious—she was telling me that my daughter Mary's baby was to be born that day!

In great relief and celebration, I rose to walk toward this elegant messenger, to embrace her with gratitude. When half the distance between us was closed, she calmly rose and trotted into the woods, with bushy tail flowing and rich markings of red, brown, and grey rippling. The next morning, I returned to take her picture, but all that remained was this white post.

And that night, Mary called, in exhausted voice, to say that half an hour previously she had given birth to a beautiful daughter, Reed Leighton Zema. Here she is, hours old!

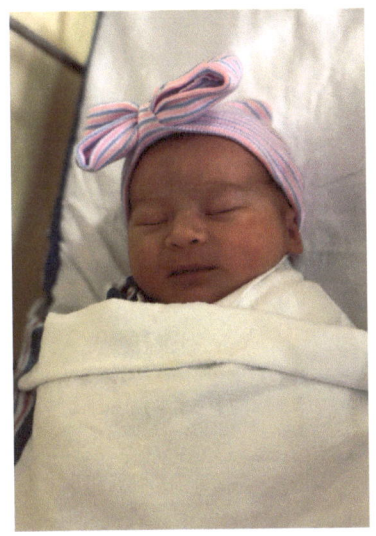

Reed Leighton is a resident of Manhattan, but we now know her spirit is also afoot in Pike County, Ohio. And she has her own indigenous name—*Coyote Blossom*, delivered by one of this land's most astute guardians.

Her parents were married here two years ago before this altar, and their spirits are imbued with this land as well.

After four days of pondering the advent of *Coyote Blossom*, I offer that experiencing one's daughter give birth is like touching the center of the earth. The power and mystery therein are unfathomably reassuring.

We are experiencing a terrible shortage of eggs, as aged hens age further and cold weather with short days set in. It appears the solution is a slow six-months away, as a new set of chicks will be arriving within weeks but won't be productive for months. Egg distribution will be awkward in the meantime. Here we are preparing the corncrib for the hens' winter lodging. We are also investing in larger mobile housing for hens to increase supply for customers. So, bear with us as we work through this issue, please!

As deer-hunting season dawns, we were blessed again this fall with the gift of a gallon of honey by a thoughtful neighbor, seeking permission to follow his prey over our property should it travel our way. The honey is thick with texture and the rich taste of minerals, though we know it comes from blossoms of plants. We savor this honey in our tea throughout the winter and consider it liquid gold.

May coyotes bring blossoms to us all, wherever we are.

Afterword
July 22, 2019

Thank you for undertaking this journey with Susan and me. We hope you have gained insight into the triumphs and struggles involved with raising and preparing grassfed meats. We also hope you feel connected to the process in general and to a local farm like ours, in particular. With this connection, a binding link is forged, reinforcing how actions of the rural directly affect the urban and vice versa.

Urban partners and customers are our lifeline, for without their weekly engagement, we quickly become moot, as the industrial food system will never support us. But we provide what it never will—handcrafted, nutrient-dense food with soul! And since it is the awakened soul that stands at the heart of successful living, sought by most, we feel well aligned for the future.

It is a struggle to raise handcrafted food at a scale that provides sufficient return of rest and profit for producers. This very struggle, however, is like a source of minerals flowing from our hearts to our food, imbuing it with unique richness transmitted to the customer. What in life that is worthwhile does not involve struggle? In the end, the best of lots is simply to be able to choose the struggle in which one is invested. And we are fortunate enough to be able to choose raising grassfed meats. In so doing, we have become deeply connected to and sincerely grateful for the opportunity to serve those with awakened souls!

For more information, visit:
www.redstonefarm.org

Credits

Front Cover: © 2016, Michael Wilson, *Drausin Wulsin with border collies*
Page 17, front cover: © 2014, Sebastien Hue, *Four cows in tall grass*
Page 76: Copyright © Terminal Mosaic Co, Mural Design: Winold Reiss, *Baldwin Piano Company*, Photography: Gregory Thorp
Page 84, 141: © 2015, Sebastien Hue, *Chili with sour cream on top*
Page 137: Copyright © 1971, 2011 by Wendell Berry, from *Farming, A Hand Book*. Reprinted by permission of Counterpoint Press.
Page 189: © 2017, Sebastien Hue, *Slider with greens*
Page 198: © 1956, The Associated Press, *Georgia Gilmore*
Page 225: © 2019, Ann Geise, *Drausin and Susan with cows*
Back Cover Aerial: © 2019, Mike Seely of Seely Portraits, Greenfield, OH
All other photographs © Drausin Wulsin

Index

Animal Impact 68, 69, 112
Annuals 112, 140
Arc of Appalachia 209, 163, 188
Ash Trees 220
Auction 30-31
Bacon 153, 167, 188
Barns 137-138
Beef Barbacoa 127
Beef Kabobs 170
Berry, Wendell 101, 137-138, 171
Biodiversity 59, 64, 124-125, 176
Bolognese Sauce 78, 84, 113, 182, 204
Bone Broth 46, 65, 119, 138-139
Brisket 97
Bulls 29-30, 33-34, 99, 190-191, 201
Butterfat 27, 29-30, 44
Carbon 51, 80, 114-116, 129, 133, 160
Cattle 13-14, 17-18, 24, 29-32, 33, 40-41, 44-45, 62, 68,
. . . . 93, 99, 106, 115, 119, 125, 140, 165-166, 170, 173,
. . . . 176-177, 179, 180, 182, 187, 190-191, 195, 202, 203, 217
Chickens 25, 34, 67, 81, 100, 210
Chili 28, 32, 78, 84, 141
Compost 55, 61, 68, 99

Corral 20, 205-206

Dairy Calves 20, 32, 36, 57, 93, 99, 106, 206

Dogs 55-56, 83, 119, 142-144, 148, 183, 189

Diet 44-46, 160

Edible Ohio magazine 191

Farm Tour 18, 128-130, 160-161, 173, 178-179, 209

Fat 44-45, 140-141

Fencing 47-48, 83

Flooding 48, 89, 107, 146-147, 203, 216, 220

Foot Rot – Sheep 89

Forest 39-40, 92, 135-136, 175-176, 220

Freezers 31-32, 204

Fried Chicken 109

Fungi 175-176

Gilmore, Georgia 198-200

Hens 138, 146, 157, 161, 186, 217, 224

Herb Garden 119-120, 173, 214

Hogs 65, 81, 167, 179, 182, 196, 217

Holding Lot 99

Horses 40, 48, 57, 59, 98

Housing 42, 106, 211-213

Hyde Park Market 156-157

Industrial Agriculture 51, 64, 73, 101-102, 115

Investors 104

Jefferson, Thomas 85, 179

Kitchen 71, 95, 106, 116-117, 120, 181, 212

Labor 66, 76-77, 91-92, 172, 200

Lamb Chops 65, 121, 145, 221

Lamb Shanks 69

Lamb Shoulder 183

Land Stewardship 22, 25, 50-51

Leg of Lamb 52-53, 63
Logging 39-40, 48
Mennonite Neighbors 18, 25, 40, 42, 86, 173, 205-207, 217
Manure 54-55, 68, 99, 166
Margin, Edge Effect 59-60
Mud 48, 194, 203
Native Artifact 111
Omelettes 139
Organic Matter 18, 50-51, 55, 68, 88, 103, 108,
. . . . 112, 116, 141, 147, 160, 165
Passion 27, 82
Parasites – Sheep 10, 171
Pastures 17-18, 20, 24-25, 61, 68-69, 99, 114,
. . . . 124-125, 160, 165-166, 175-176, 182
Poached Eggs 93, 161
Pork Chops 81, 90, 131, 136, 177, 186, 197, 207
Pork Shank 134
Pork Shoulder 87, 200
Pork Tenderloin 157
Pot Roast 71, 95
Red Devons 13, 29-30, 33-34, 93, 99, 119, 140, 190
Rams 75, 89, 190
Rest 16, 22-23, 61-62, 69, 197, 225
Roadways 48, 168-169
Roast Chicken 34-35, 38, 43, 67, 75, 174, 196, 210
Savory Institute 54, 79-80
Science vs. Nature 73-74
Serpent Mound 162
Sheep 18, 61-62, 68, 80, 83, 89, 125, 142,
. . . . 147-148, 160, 171, 182, 200, 208, 216-217
Shortrib Burgers 117, 177, 189, 196, 204
Sliders 10, 28, 141, 163

Slider Shack..163
Steaks........... 20-21, 30, 41, 58, 60, 104, 123, 140-141, 148, 180, 202
Teicholz, Nina...46
Vision.................................... 38, 55, 82, 133, 169
Water Management........... 48, 105-108, 146-147, 175, 192-194, 203
Waterlines...........................36-37, 56-57, 106-107, 147
Wetlands............11, 24, 33, 39, 63, 90, 96, 107, 118, 132-133, 209
White Oak Pastures....................................79, 81
Wildfire...135-136
Wildlife................61, 83, 133, 143, 159-160, 171, 177, 180, 222
Wildflowers....................... 63, 101, 103, 116, 152-153, 163
World Health Organization..............................73-74

www.ingramcontent.com/pod-product-compliance
Lightning Source LLC
Chambersburg PA
CBHW042042290426
44109CB00001B/5